P9-DGM-405

The Geography of the Internet Industry

The Information Age Series
Series Editor: Manuel Castells

There is a growing interest in the general audience, as well as in universities around the world, on the relationships between information technology and economic, social, geographic, and political change. Indeed, these new relationships are transforming our social, economic, and cultural landscape. Social sciences are called upon to understand this emerging society. Yet, to be up to the task social sciences must renew themselves, in their analytical tools and in their research topics, while preserving their scholarly quality.

The Information Age series is the "Nasdaq" of the social sciences – the series that introduces the topics, the findings, and many of the authors that are redefining the field. The books cover a variety of disciplines: geography, sociology, anthropology, economics, political science, history, philosophy, information sciences, communication. They are grounded on original, rigorous research and present what we really know about the Information Age.

Together, the books in *The Information Age* series aim at marking a turn in the academic literature on information technology and society.

Published

Work in the New Economy: Flexible Labor Markets in Silicon Valley
Chris Benner

Bridging the Digital Divide
Lisa J. Servon

The Internet in Everyday Life
Barry Wellman and Caroline Haythornthwaite

The Geography of the Internet Industry: Venture Capital, Dot-coms, and Local Knowledge
Matthew A. Zook

The Geography of the Internet Industry

Venture Capital, Dot-coms, and Local Knowledge

Matthew A. Zook

Blackwell Publishing

BLACKWELL PUBLISHING
350 Main Street, Malden, MA 02148-5020, USA
9600 Garsington Road, Oxford OX4 2DQ, UK
550 Swanston Street, Carlton, Victoria 3053, Australia

The right of Matthew A. Zook to be identified as the Author of this Work has been asserted in accordance with the UK Copyright, Designs, and Patents Act 1988.

First published 2005 by Blackwell Publishing Ltd

1 2005

Library of Congress Cataloging-in-Publication Data
Zook, A. Matthew.
   The geography of the Internet industry : venture capital, dot-coms, and local knowledge / by Matthew A. Zook.
       p.   cm. – (The information age series)
   Includes bibliographical references and index.
   ISBN 0-631-23331-8 (hardback : alk. paper) – ISBN 0-631-23332-6 (pbk. : alk. paper)
   1. Internet industry – Location.   I. Title.   II. Series.

   HD9696.8.A2Z66 2005
   338.4'7004678'0223–dc22

                                                                    2004019040

ISBN-13: 978-0-631-23331-2 (hardback : alk. paper) –
ISBN-13: 978-0-631-23332-9 (pbk. : alk. paper)

A catalogue record for this title is available from the British Library.

Set in 10.5 on 12.5 pt Palatino
by SNP Best-set Typesetter Ltd, Hong Kong
Printed and bound in the United Kingdom
by TJ International Ltd, Padstow, Cornwall

The publisher's policy is to use permanent paper from mills that operate a sustainable forestry policy, and which has been manufactured from pulp processed using acid-free and elementary chlorine-free practices. Furthermore, the publisher ensures that the text paper and cover board used have met acceptable environmental accreditation standards.

For further information on
Blackwell Publishing, visit our website:
www.blackwellpublishing.com

Contents

Figures

Tables

Maps

Series Editor's Preface

The economy of the Information Age is not placeless, in contrast with the superficial predictions of futurologists. The production of information and knowledge is in fact rooted in specific places that Peter Hall and myself theorized as milieus of innovation years ago. The Internet has a geography, and the geographic location of Internet domains is one of the most spatially concentrated location patterns. The geography of an Internet-based economy and society is made of nodes and networks that criss-cross the planet. Thus, it is neither spatial dispersion nor spatial concentration that characterizes the new geography but the interaction between both processes, what I have named the "space of flows."

Our knowledge of the geography of the Internet has benefited a great deal from the decisive contribution of Matthew Zook's pioneering research. Although a number of scholars have worked in this field for some time, as Zook points out in his careful list of bibliographic references, in my personal assessment the study by Zook is the most complete empirical analysis to date of the spatial patterning of Internet-based production of information. He developed, years ago, a statistical mapping of a representative sample of Internet domains worldwide, and kept updating this sample, catching up with the speed of development of the Internet (I must say he was probably helped by the recent slowdown of Internet diffusion). He thus showed the high level of concentration of Internet domains by country, by region, by metropolitan area, and even by specific locations within metropolitan areas. He showed that the production of Internet content closely follows the geography of information and knowledge. But he went beyond that, explaining the formation of some of the highest nodes of Internet-based activities, including the San Francisco Bay area, through careful case studies and in-depth interviewing. He argued, with solid data in hand, that the location of venture capital firms has a very strong influence on the development of Internet innovation and Internet-based production of information. Should we accept this analysis, as I do, there are extraordinary consequences for regional and local development policies. Financial institutions of innovation are probably more important for economic growth in this knowledge economy than the location of research universities.

The importance of Zook's work goes beyond the substance of his findings. It is the style of his research that brings innovation to the field of social sciences. He moves freely across disciplinary boundaries, as one should do in dealing with the analysis of information technology-related processes, since this is a transversal phenomenon that affects every domain of society. He also mixes, always with rigor and scholarly care, various methodologies and traditions of inquiry, statistical analysis as well as interviewing, computerized geographic techniques, and documentary work. He also knows, and combines, various relevant theoretical frameworks, escaping from the iron cage of a nonexisting unified theory. He challenges established knowledge, but knows the research and thinking that preceded his. He is a representative of a new generation of young scholars, ready to study and understand our new economy, and our new geography, in continuity with the best tradition of social sciences, opening new ground when it becomes necessary to do so. As with all innovators, his work does not fit easily in one academic field, but it connects geography, and the study of spatial transformation, to the analysis of the new economy, to technologic change, to the institutional environment of innovation, and to the dynamics of producing and distributing knowledge and information.

Matthew Zook's book is the first systematic assessment of the relationship between the Internet and the geographic dimension of the network society. It also proposes a new style of research, and blends existing theories of the geography of innovation in an original analytic framework. I am convinced that reading, and critique, of this book will contribute considerably to our understanding of processes of local and regional development, and to our ability to act upon them.

Manuel Castells
Barcelona/Los Angeles
July 2004

Acknowledgments

Although the process of writing a book is largely solitary, I marvel at the number and range of people to whom I am indebted. From the first months in Berkeley, California, to the final edits made in Lexington, Kentucky, I have relied upon those around for input, critique, and support. Without such help, this book could not have been completed.

My thanks go first to Manuel Castells for including my work in *The Information Age* series. His scholarship and mentoring have been instrumental in shaping my conceptualizations of the information age and research agenda. I greatly appreciate the time and energy that he has offered, and I look forward to passing his influence to future students. I am also profoundly grateful to my other advisors at Berkeley whose invaluable training and guidance made this book possible. In particular, AnnaLee Saxenian chaired my dissertation, introduced me to the exciting arena of high technology and regional development, and provided me with a foundation for future projects. Dick Walker engaged me with wonderfully probing questions on the theoretical implications of this work and helped me to place it in context with economic geography. I offer my deepest thanks to all of you.

I further acknowledge the tremendous benefit of input and interaction with fellow researchers at Berkeley and elsewhere. Thanks to the global regions group, Ted Egan, Sean O'Riain, Balaji Parthasarathy, and Gerald Autler for wide-ranging discussions. Many thanks to Gary Fields, Peter Hall, and Larissa Mueller, who gave invaluable feedback, even in the early and highly disorganized stages. Similarly, I am indebted to Anthony Townsend, Martin Dodge, and Sean Gorman, my fellow venturers into the geography of the Internet, with whom I could discuss the finer point of domain name versus host counts versus bandwidth to my heart's content. I am grateful to Karen Chapple, Yuko Aoyama, and John Thomas who provided vital suggestions on both my research and personal life. In addition, I am grateful to all my "bowling league" buddies who provided important feedback and support.

Additional thanks go to Mike Teitz and the Public Policy Institute of California for their key support of my research on the dot-com bust. My work is unquestionably richer for my ability to explore the repercussions in the San Francisco Bay technology community, post 2000. Most recently, my colleagues in the Geography Department at the University of Kentucky have welcomed me generously, providing a home for me and the manuscript that has become this book.

A number of friends and family have generously contributed their time and homes to my research. Special thanks to Chris and Alison Ney, Mike Johnson and Greg Gould, Dena Beltzer, Larissa Mueller, and Doug Webster who allowed use of their apartments and office space as I conducted fieldwork and writing. I also wish to thank the people who gave generously of their time to be interviewed. This book would not have the same depth if not for your insights and sharing of experiences. I wish to thank my entire family, especially my parents Gordon and Bonnie Zook, for raising me to be intellectually curious and encouraging me to pursue education. Special thanks to my daughter, Maara, who constantly reminds me what life's real priorities are. And finally, all my love and heartfelt thanks to my wife, Eva Ensmann, for being my true partner in life and work. This book (which you have experienced from its earliest to latest versions) is dedicated to you.

Map 2.1, ARPANET 1971, and Map 2.2, ARPANET 1980, are from Internet Archive's ARPANET paper collection http://www.archive. org/texts/arpanet.php.

Parts of this book have previously been published in solely authored journal articles, and I should like to acknowledge the following for permission to reproduce this material: versions of Map 3.6 and Map 3.7 first appeared in M. A. Zook (2000) "The web of production: the economic geography of commercial Internet content production in the United States," *Environment and Planning A*, 32: 411–26, permission granted by Pion Ltd, London; sections of chapters 4 and 6 are based on edited extracts of M. A. Zook (2004) "The knowledge brokers: venture capitalists, tacit knowledge and regional development," *International Journal of Urban and Regional Research*, 28(3): 621–41, permission granted by Blackwell Publishing; and sections of chapter 5 are based on M. A. Zook (2002) "Grounded capital: venture financing and the geography of the internet industry, 1994–2000," *Journal of Economic Geography*, 2(2): 151–77, permission granted by Oxford University Press.

1

Uncovering the Geography of the Internet Industry

The Internet has revolutionized the way the world communicates. In less than a decade (see figure 1.1) it has transformed from a relatively obscure computer network into a global system of hundreds of millions of networked computers (hosts) and tens of millions of formal sites for interaction and commerce (domains). Contacting someone on the other side of the world is as simple as a mouse click and billions of web pages offer a cornucopia of content, commerce, interaction, services, and products. Paralleling the expansion of the size of the Internet were the fervent efforts by individuals and companies to harness the perceived power of the growing network for personal enrichment and commercial gain. The activity surrounding these efforts was extraordinary as measured by any number of variables, including media attention and stock-market investing. In short, the Internet at the *fin de siècle* represents a time of historic change and frantic endeavors to establish footholds in this new medium.

Particularly intense were the energies and capital directed toward the dot-com companies which made up the Internet industry. The factors and dynamics behind the creation, clustering, and retrenchment of this new industry from 1994 to 2003 is the focus of this book. Beginning with the founding of Netscape Communications in April 1994 and extending through the market downturn in April 2000, it was a time of big plans, loose capital, and hot hyperbole. Companies frenziedly pursued a variety of new business models designed to make them the ascendant corporations of the 21st century. The world was changing and everyone wanted to be at the center of it.

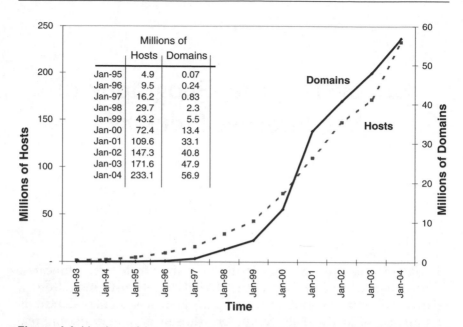

	Hosts	Domains
Jan-95	4.9	0.07
Jan-96	9.5	0.24
Jan-97	16.2	0.83
Jan-98	29.7	2.3
Jan-99	43.2	5.5
Jan-00	72.4	13.4
Jan-01	109.6	33.1
Jan-02	147.3	40.8
Jan-03	171.6	47.9
Jan-04	233.1	56.9

Figure 1.1 Number of Internet hosts and domains (generic top-level domains and country code top-level domains), July 1992 to January 2004.
Source: host counts based on data from the Internet Software Consortium (http://www.isc.org/); domain counts 1992–98 (Zakon, 1999), 1998–2004 (author's survey).

The market downturn beginning in 2000 showed that the grandiose expectations associated with most of these dot-com business plans were simply not going to come to pass. Companies that eagerly pursued the dot-com moniker in the late 1990s found themselves struggling for survival and in many cases simply disappeared. However, despite very real negative economic consequences of this shakeout, an Internet industry backed by venture capital did coalesce during this time. Multibillion dollar companies such as Yahoo!, eBay, and Google remain central to how the Internet is used worldwide and demonstrate that this industry is not simply smoke and mirrors but a continuation of a historical and geographic process of technological and economic development.

The Persistence of Geography

This story also shows the fundamentally geographic nature of the development of the Internet and contrasts sharply with commonly held assumptions that physical locations would become irrelevant. In the mid-1990s pundits predicted the "death of distance" and the "end of cities" and confidently envisaged a world where social and economic interactions would increasingly take place in virtual space. As the 20th century came to a close, however, the rhetoric of "spacelessness" became increasing difficult to reconcile with reality, particularly within the heavily clustered Internet industry. As this book documents in its primary case study, the milieu of the San Francisco Bay region was and is a key location for the Internet industry and the companies that form it.

The reasons behind this paradoxical clustering of a "placeless" industry are tied to the fact that the creation of successful companies depends not simply upon a supply of business plans, skilled labor, infrastructure, or capital, but also relies on the way in which these resources are marshaled and organized. Ironically, precisely because the Internet made certain types of information more widely available, regional environments that facilitated the creation, organization, and use of unique knowledge were central in the development of the industry. Equally paradoxical, a key mechanism behind the clustering of this so-called "placeless" industry was capital investing.

Capital is often perceived as freely flowing to the location of the greatest opportunity for return, but the venture capital investing that was central to the Internet industry was much more than simply money. As Martin (1999, p. 11) argues, "money is not just an economic entity, a store of value, a means of exchange or even a 'commodity' traded and speculated in for its own sake; it is also a *social relation*." Many venture capitalists have strong local orientations when seeking portfolio companies in order to maximize their key tool in risk management, i.e., unique knowledge about new technologies, entrepreneurs, and competitors' actions. Venture capitalists rely upon this knowledge, built up through social and professional interaction, to make investments in situations of great uncertainty.

Therefore, venture capitalists can be characterized as knowledge brokers who acquire and create intelligence through personal (and generally local) networks about industries, market conditions, entrepreneurs, and companies through a constant process of interaction and observation. While capital in the most general sense of the word, i.e., money, provided the fuel for many Internet companies, it was the transmission and use of tacit (noncodified) knowledge that in many ways was more valuable. The ability of venture capital to quickly supply this type of value-added input is dependent upon the quality of its networks and is greatly assisted by geographic proximity, which in turn contributed to the clustering of dot-com firms.

However, the knowledge and local networks created and used by venture capitalists do not emerge overnight. Rather venture capital systems develop alongside and concurrent with the industrialization and development process. Crucial to the operation of these regional financing systems are the feedback loops that emerge over time as venture capitalists, entrepreneurs, and labor come together in various new ventures. Even if the new firms do not succeed, valuable information, experience, and contacts develop during the process. These new or strengthened connections within a regional system provide the basis for subsequent efforts to form innovative firms. The case of the Internet industry illustrates the advantage that accrues to firms and regions with the ability to move and adapt quickly to new innovations. In particular, the San Francisco Bay experience demonstrates how regional venture capital systems are built through a process of incremental steps that lay the foundation for subsequent rounds. As a result, firms within the region (such as Yahoo! or eBay) were able to move quickly when the opportunity of the commercial Internet emerged in the mid-1990s.

This advantage also had its downside as the initial wave of investing and new company formation turned into a frenzy of money chasing bad business models. The great advantage of venture capital investing at the start of the era, i.e., access to unique knowledge to select technologies and firms, was diluted in a wave of bloated capital funds, inflated and copycat investing, and a preoccupation to "get big fast" at any cost. In most cases, investment decisions were individually rational but built upon irrational expectations surrounding the promise of a new technology. The end result was a large influx of capital without much oversight or direction. Money was spent, market share was garnered, publicity was gathered but despite these temporary successes, many dot-com companies were unable to transition into lasting business models.

Precisely because the San Francisco Bay region was a center for the early Internet industry, it was also ground zero for this later period, albeit with negative results in terms of relevance and longevity of the new firms. However, as tempting as it may be to stereotype the dot-com era as 20-something chief executive officers (CEOs) wasting millions of dollars on Superbowl ads, expensive office chairs, fussball tables, and parties, the impact of dot-com boom and bust has much more complex implications in the short and long term. Even in the face of numerous bankruptcies, accounting scandals, and a weak economy, the dot-com era is not without its upside. Moreover, the rise of dot-com firms is not so much an anomaly but the most recent manifestation of Schumpeterian creative destruction.

Thus, despite telecommunications technologies and global capital flows that have vastly expanded the geographic range of economic interaction, regional milieus remain central to economic development in the 21st century. The development of the Internet industry is fundamentally embedded in geography and defies simple expectations of diffusion and the demise of cities and instead illustrates the continued importance of particular regional and urban nodes in an increasingly globalized economy. It is, however, neither a short-term nor straightforward process to create the conditions for innovative regional development. As the Internet industry shows, simply injecting money indiscriminately can lead to ill-advised investments and short-lived companies. Nevertheless, the ability to adapt to the changing dynamics of the economy will continue to be relevant in the future as regions attempt to reinvent their economies, enter new industries, and innovate.

Defining the Internet Industry

The decision of what to include in the working definition of the Internet industry (referred to interchangeably as the dot-com industry) is difficult. Although an instantly recognizable and widely used term, it cannot be easily reduced to a specific sector, business model, or firm type. In fact, at the most basic level, it is simply an indication that a company uses the Internet in some form. As the use of the Internet by businesses becomes increasingly common, the distinction of being an Internet-using company has begun to have as much significance (or lack thereof) as being a phone-using or fax-using firm. In short, the Internet has become an essential part of conducting business in the USA and the world.

Despite this imprecision, the term "Internet/dot-com firm" invokes a certain kind of enterprise that emerged in the closing years of the 20th century when companies first began experimenting with the Internet as a part of business. The promise of the Internet in the mid-1990s was so compelling that people confidently predicted the wholesale transformation of sectors as diverse as grocery retailing and the purchase of steel and chemicals. While these businesses are continuing to evolve with their use of the Internet, the immediate changes hoped for by dot-com companies and their investors were not forthcoming. Today, with the increasingly widespread use of the Internet it is more problematic to refer to companies as Internet companies simply because they use the Internet. Nevertheless, for the period of time examined by this book it remains a useful term.

In practice, this book defines its object of study on the basis of three interlinking criteria. The first is the possession of a business model that was primarily Internet based and/or whose operation would not be possible without the Internet. A majority of these companies were founded or completely restructured between 1994 and 2000 with the Internet as a central component to business. These business models could include any number of foci, e-commerce, content generation, advertising, community, or information services and be oriented toward consumers, businesses, or government both locally and world-wide. While this definition encompasses a wide range of companies, it accurately reflects the enormous range of experimentation taking place during the closing years of the 1990s.

The second criterion for inclusion as an Internet firm is the expectation of extraordinarily fast growth through the creation of new markets or the disintermediation of existing markets and value chains. While in retrospect these expectations seem unreasonable, conventional wisdom at the time within the business community was that dot-com companies were poised to reinvent and dominate their markets.[1] This expectation also led to a reliance on nontraditional metrics such as growth in users rather than profitability for evaluating these companies. Dominating a changing market quickly became many dot-com companies' primary goal and was pursued with little regard to cost and with the full support of investors.

This potential for fast growth raises the third and final criterion for inclusion as a dot-com company, i.e., financial backing from risk investors interested in high returns. Often referred to in the early stages as venture capital, this type of financing encompasses a much wider range, from individual seed investments by the entrepreneur, family, and friends, investments made by corporations in spun-out divisions, formal venture rounds by limited partnership venture capitalists, to initial public offerings (IPOs) oversubscribed by institutional and small investors around the world. At the height of the boom it also included millions of small investors worldwide using discount online brokerages to secure a piece of the dream.

In short, Internet companies were young, fast-growing, risk-capital backed companies which used the Internet as an integral part of their business model. While any number of companies cross these definitional boundaries, this is the essence of the Internet industry and represents the object of this book's research.[2]

Plan for the Book

In order to analyze the role of geography in the development of the Internet industry it is first necessary to describe its spatial distribution. The recent and nonhierarchical nature of the Internet, however, makes studying its geography difficult. Because data long did not and for the most part still does not exist, this book spends considerable time describing and mapping the Internet, particularly its commercial aspects. Chapter 2 briefly sketches the history of the precommercial Internet and describes the contours of the global geography of its use. Chapter 3 considers the specific geography of the Internet industry and through a number of indicators triangulates its concentration and clustering. Together these two chapters outline the unique geography of the Internet industry, the phenomenon the rest of the book is devoted to explaining.

Chapter 4 lays out a theoretical argument building upon earlier research on the role of networks, institutions, and conventions within a regional context. In particular, the provision of financing for firms, which has been a relatively overlooked factor in firm formation and regional development, was key in the case of the Internet industry. Chapter 5 substantiates this theoretical argument with an analysis of the amount and destination of venture capital investments and shows that the clustering pattern of the Internet industry is closely tied to venture capital investing. However, this finding should not be taken as a pure supply-side argument in which simple access to capital equates with entrepreneurial success. Rather, as Chapter 6 develops, venture capitalists' intricate connections to regional knowledge, labor, and industries are what allowed it to play a central role in producing the key firms in the emerging Internet industry.

The experience of the San Francisco Bay region outlined in Chapter 7 provides the exemplar of the value and use of knowledge by venture capitalists. The initial period of Internet commercialization illustrates the advantage that accrues to firms and regions with the preexisting ability to move and adapt quickly to new innovations. In particular, it demonstrates how regional venture capital systems built through earlier industries (such as semiconductors, personal computers, and networking) laid the institutional foundation exploited by early

Internet firms. However, this early knowledge advantage did not prevent later excessive investment in Internet companies with extremely dubious business plans and little prospect for viability.

Chapter 8 assesses the factors and actors that led to the boom of dot-com firms within the San Francisco Bay area outlined in the previous chapter. While in retrospect much of the activity surrounding dot-com firms was wasteful and ill-planned, there were several compelling factors behind the willingness of normally rational people to accept much of the "irrational exuberance" surrounding the dot-com economy. The chapter highlights the widespread faith in the transformative nature of the Internet and the avarice and ambition of those involved in dot-com firms. By 2000, the Internet industry had evolved into a caricature of venture capital investing, i.e., putting as much money into play as possible with little of the value-added that traditionally accompanied it. The resulting downturn, including layoffs, bankruptcies, and a collapsing stock market, is outlined and analyzed.

Chapter 9 extends this history but rather than simply concentrating on the well-known story of the decline of dot-com companies, it presents a contrarian view of the dot-com decline and demonstrates some of the ways the dot-com boom (and even the bust) has helped develop the San Francisco Bay economy. The chapter argues that the Internet industry is strongly linked to the processes of innovation and creative destruction that have been central to capitalist development for centuries. Firms are founded, grow, and disappear with great regularity. Earlier waves of innovation also exhibited overinvestment and new company formation beyond what market forces could reasonably be expected to support. While the dot-com era was the most recent and a particularly spectacular manifestation of this ongoing process, it was not a fundamental departure from economic development patterns both specifically for the San Francisco Bay region and more generally.

With 20/20 hindsight, much of the rhetoric of the late 1990s concerning the potential for Internet startups to "destroy geography" or challenge existing "offline" companies seems excessively naive. Although the dubious nature of many dot-com business plans is undeniable and it is clear that speculation drove much of the investment activity, there are nevertheless longer-term implications for regional development. At the most basic level this includes the creation of real and lasting multibillion dollar companies, while more nuanced analysis points to the continued relevance of geography and the role of financiers in transmitting knowledge. Although not as obvious as overnight millionaires and flashy companies, these factors remain central to economic development in the 21st century. Despite the reversal of fortunes of many Internet firms and the related fallout in terms of unemployment, the capability to create and use knowledge exhibited by venture capitalists remains relevant for future regional economic development.

2

Origins and Shape of the Internet

The Internet has captured the world's imagination by its ability to distribute information on a real-time basis across the globe. Growing from an experimental computer network, it has transformed the way the world communicates and provides the appearance of a system without geography. However, the development and use (particularly the commercialization) of the Internet has always had clear connections to specific places and people. This is reflected in both the creation of the Mosaic web browser at the University of Illinois and the dominance of the San Francisco Bay region during the dot-com boom, both of which were among the first 15 wired locations.

While this suggests that places with longer histories of involvement with the Internet were at an advantage, the geography of the Internet remains an evolving phenomenon. Long the exclusive domain of computer scientists and academics, the invention of the World Wide Web and graphical browser transformed the Internet into a global and increasingly commercial medium. The ease with which users could connect to remote locations provided many with a sense that cyberspace was creating a placeless forum for interaction and commerce. Despite this appearance, the Internet – whether the original 15 nodes of the ARPANET or the 233 million computers connected to it in January 2004 – cannot exist without the fiberoptics, routers, and computers that distribute its packets and most importantly the people who create and consume its content.

History and Early Geography of the Internet[1]

The Internet's origin is based in the US Department of Defense and its Advanced Research Project Agency (ARPA). Charged with keeping the USA ahead of the Soviet Union in terms of military technology, ARPA pursued advanced research in a number of fields including computer science and telecommunications. In 1962 ARPA formed the Information Processing Techniques Office (IPTO) that funded computer science research and would eventually provide the impetus for developing the computer networks between ARPA-funded universities that were the precursor of the Internet (Norberg, O'Neill, and Freedman, 1996).

Around the same time that IPTO was founded, two researchers, Paul Baran in the USA and Donald Davies in England, independently developed the idea of "packet switching," the fundamental technology underpinning the Internet. Packet switching is based on the concept of dividing data into smaller units of a standard size, i.e., packets, which can then be routed independently to their final destination where they are reassembled. This method allows for more flexibility in routing data and greatly increases the amount of traffic that can be transmitted across communication lines. Baran saw packet switching as a way to improve the survivability of military communications while Davies's main concern was on interactive computing, i.e., increasing the number of users who could access expensive mainframe computers (Abbate, 1999). Eventually Lawrence Roberts, the manager of ARPA's effort to network the universities it funded, decided to utilize packet switching in the project.

The ARPANET project first connected four nodes in the West (UCLA, SRI, UCSB, and University of Utah) and then extended the system to 15 computer science centers funded by IPTO. As map 2.1 illustrates, the geography of the earliest precursor to the Internet foreshadowed future concentrations in the San Francisco Bay, southern California, and New England. In 1972 Lawrence Roberts left ARPA and Robert Kahn became the director of ARPANET. Kahn was particularly interested in how radio and satellite communications could be used for transmitting data because of its potentially important military applications. The existence of three different systems for transferring packet data (landlines, radio, and satellite) presented Kahn with the problem of how to connect them to one another. Although each system was built around packet switching, they used very different transmission protocols.

The problem of connecting these systems led Kahn and Vinton Cerf to devise a set of protocols, i.e., transmission control protocol (TCP)/Internet protocol (IP), that formed the basis for "networking networks" which would eventually become the underlying protocols

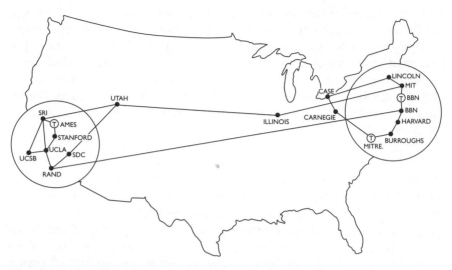

Map 2.1 ARPANET 1971.

for the Internet (Abbate, 1999, p. 122). Cerf and Kahn actively solicited the involvement of a number of other researchers in the creation of this protocol, resulting in a design that was readily accepted by many. The first demonstration of this internetworking capability took place in 1977 and joined the three packet switching networks of ARPANET. Eventually the managers of ARPANET would require that all nodes use the TCP/IP protocol, making it a de facto and long-lasting standard.

During the late 1970s and early 1980s there was great interest on the part of computer science departments at non-ARPA-funded campuses in gaining access to the network. The result of these efforts included the founding of CSNET in 1981 by the National Science Foundation (NSF) to provide a network between computer science departments and the creation of BITNET, funded by IBM and designed to provide electronic mail capabilities for academic communities in general (Hart, Reed, and Bar, 1992, p. 670). These additional networks and the introduction of the personal computer (PC) and local area networks (LANs) greatly increased the number of connected computers. This resulted in extremely rapid growth, e.g., in 1982 only 15 networks were connected to ARPANET but by 1986 there were more than 400 (Abbate, 1999, p. 188).

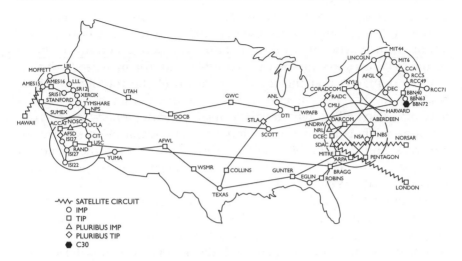

Map 2.2 ARPANET 1980. Note that this map does not show ARPA's experimental satellite connections. Names shown are IMP names, not (necessarily) host names.

As map 2.2 illustrates, the network expanded to include new sites in the Midwest and South, although California and New England continued to dominate the ARPANET structure. Equally significant for the later geography of the Internet is the early connection to the UK that was and continues to be among the largest non-US concentrations of Internet activity.

In the mid-1980s, the NSF began to establish supercomputing centers around the country. Due to the expense of this equipment, only five of these centers (Cornell, Princeton, Pittsburgh, University of Illinois at Champaign–Urbana, and University of California at San Diego) were initially funded. In order to allow researchers at other locations to connect to these supercomputers, the NSF also funded the development of a central backbone of high-speed lines known as NSFNET. NSFNET eventually came to use the TCP/IP protocol and collaborate with ARPANET in sharing some backbone connections and operating costs. The working relationship between the two networks meant that it was a relatively straightforward decision and process to switch the ARPA hosts to NSFNET when ARPANET was finally shut down on February 28, 1989 (Hart et al., 1992).

By 1990, traffic on the NSFNET had grown tremendously and there was increasing pressure from companies interested in using the NSFNET for private purposes. Because NSFNET was publicly funded, all use of it was governed by the NSF's "acceptable use" policy that in theory limited all uses to research and instruction at universities, although in practice this rule was hard to enforce.[2] Interest in more commercial uses for the Internet, coupled with the increased administrative demands that the rapid expansion of the Internet entailed, resulted in the NSF's eventual disengagement from managing the network. The NSF slowly devolved itself of various network responsibilities, such as handing over domain name registration services to Network Solutions in 1993, but continued to manage the principal backbones until 1995 when this task was taken over by private networks (Abbate, 1999, pp. 196–9).

Although the number of people connected to the Internet grew exponentially during the late 1980s and early 1990s, it still remained largely outside the mainstream. Command line interfaces, lack of search capabilities or indexes, and the primarily computer-centric focus of much of its content all contributed to it being the domain of technophiles. The introduction of Archie, Gopher, and Veronica search software in 1990, 1991, and 1992 made finding things on the Internet easier, but they were still a far cry from the graphical window interfaces available on PCs. It was not until the development of the World Wide Web and the browser that the stage was set for the rapid introduction of the general public to the Internet.

The principal person behind the creation of the World Wide Web was Tim Berners-Lee, a British physicist and computer programmer who developed the World Wide Web protocols and browser while he was working at the European Laboratory for Particle Physics (CERN) in Geneva.[3] He first began working on the issue of organizing information through user-defined links when he was employed at CERN for six months in 1980.[4] Upon his return to CERN in 1989, he again confronted the problem of tracking information but this time began to think of it at the organizational level.[5] In response to this problem, Berners-Lee proposed the construction of a hyperlinked information system, originally named MESH, that allowed people to track the numerous project and personnel involved with CERN (Berners-Lee, 1989).

A year and a half later, Berners-Lee received CERN's agreement to fund the project, now renamed the World Wide Web, for six months (Moschovitis, 1999). In November 1990, Berners-Lee wrote the first version of the browser and web-server programs and demonstrated them at CERN in December. The software was distributed to all CERN users on May 17, 1991 and to the general Internet public in late 1991 (Naughton, 2000, p. 235). As CERN funding for the World Wide Web project ended, Berners-Lee encouraged the Internet community to con-tinue to write software for the World Wide Web.[6] A number of new browsers such as Erwise, Midas, Cello, and Viola appeared and were distributed free of charge from CERN's web servers (Naughton, 2000, p. 235).

Despite the impressive accomplishment of creating a coordinated system of information retrieval, the protocols created by Berners-Lee did not alter the fundamental structure of the Internet. It was possible to use other types of protocols such as ftp or Gopher to acquire the same data accessed via a web browser. But as Berners-Lee notes, "It was basically technically trivial to go and get it. It just happened that you had to be a guru of the highest degree to actually be able to navi-gate all the networks and figure out all the programs that you would come across on your way and know what commands to give them to actually get the data back" (Segaller, 1998, p. 289). In other words, the World Wide Web was central to transforming the Internet from the domain of computer hackers to a mainstream communications medium.

By the end of 1992 a number of new interfaces and search tools such as Gopher and WAIS were being used to navigate the Internet. Although these tools represented progress in organizing, accessing, and using the Internet, they were incompatible with one another and limited in what they could do. The World Wide Web and its accom-panying protocols offered an easier interface to the Internet and in so doing opened the doors for its mass use. However, the Internet and World Wide Web of 1992 were still not particularly attractive places for most people. Most of the software only ran on high-end UNIX workstations requiring extensive configuration and it was difficult and time-consuming to find things without good indexes. Although the World Wide Web and other tools were attracting more people from the nontechnophile community, there were still only 26 web servers in existence by November 1992 (Naughton, 2000, p. 239). Other models for computer networks such as the private networks of America Online and Prodigy existed apart from the Internet and were easier to use and more inviting to the average person than the Internet and World Wide Web of the early 1990s.

At this time a University of Illinois undergraduate named Marc Andreessen started work as a student programmer at the National Center for Supercomputing Applications (NCSA) (Reid, 1997). Andreessen first encountered the World Wide Web at NCSA and decided to write a better browser that would add graphical capability to the Web. By late 1992 he and Eric Bina as well as a number of other programmers created a software program named Mosaic that was released to the Internet public in February 1993. The first version of Mosaic allowed users to follow hyperlinks by simply clicking on them with a mouse, had pull-down menus and scroll bars, and was able to load images (Reid, 1997). These features made Mosaic the most visually interactive and easy-to-use web browser to date and proved immensely popular within the Internet community, paving the way for the dot-com boom of the late 1990s.

While the appearance of Mosaic at one of the original ARPANET nodes suggests that these regions had a head start in the commercialization of the Internet, the 1990s shows that having a head start did not guarantee a concentration. Shortly after the release of Mosaic, Andreessen and the rest of the development team were recruited by Jim Clark to relocate to Silicon Valley and help found Netscape. As developed in greater detail later in the book, the geography of the Internet industry drew upon factors of technological prowess and economic organization.

Geography of Internet Users

It is clear by any account that the Internet has become a global system in which it is possible to "connect" even in some of the most remote locations. Yet "connected" is a tricky word that conveys a range of meanings, from simply accessing data or sending email to the ability to run an online business. In its broadest sense it is evident that the number of Internet users continues to grow quickly and NUA (2002) estimates that in September 2002 there were 601 million people online worldwide. Estimates at the country level or lower are more difficult to obtain but using NUA's (2002) compilation of Internet user surveys from around the globe, it is possible to assemble rough estimates of the number of Internet users for 150 countries. Because these surveys were conducted with different methodology and at different times they should not be compared too closely. Unfortunately, no other source for statistics with comparative global coverage or at a more disaggregated level is available.

The USA leads the world with 165.75 million users, followed by Japan with 56 million, China with 45.8 million, the UK with 34.3 million, and Germany with 32.1 million. The distribution of the world's Internet users and the percentage of a country's population online is shown in map 2.3.[7] Not unexpectedly, the industrialized

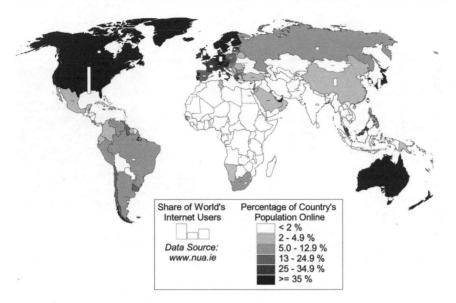

Map 2.3 Internet users worldwide, September 2002. Data from http://www.nua.ie

countries of North America, western Europe, and Japan are well represented both in terms of numbers and percent online. Moreover, there are many parts of the industrializing world such as most of sub-Saharan Africa, Southwest Asia (with the exception of the UAE), and Central Asia that have extremely low Internet usage while other countries, most notably China, have large and quickly growing Internet populations.

This pattern shows the continuing diffusion of Internet usage worldwide, albeit in a highly selective and geographically specific manner. While the USA remains the largest concentration of Internet users, this represents a decided shift from the early 1990s when almost all Internet users were located there. Figure 2.1 illustrates that the US share had shrunk to 69.1 percent in January 1997 and 27.6 percent in August 2002.

Global Geography of the Internet Industry

However, the growth and diffusion in users is by definition an indicator of the consumption side of the Internet. While it is useful in track-

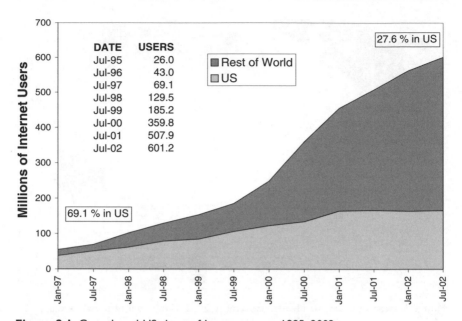

Figure 2.1 Growth and US share of Internet users, 1995–2002.
Source: http://www.nua.ie/surveys/how_many_online/index.html. Data for the number of US users prior to January 1997 is not available.

ing the growth of potential demand incentives for the Internet industry, it does not assist in understanding where the industry itself is locating. Other research (Zook, 2001) demonstrates that the supply and consumption of Internet information and products is unevenly distributed globally, with some countries "exporting" their Internet content and commerce to the world.

Obtaining accurate and meaningful geographic measures of the supply of commercial Internet content and products is a difficult undertaking. Domain names, i.e., yahoo.com or nokia.fi, are arguably the best indicator of the supply of Internet content, services, and commerce because they suggest an effort to organize and distribute some body of information and contain the unique contact information of the person or entity that registered them.[8]

Although the largest concentration of domain names remains in the USA, the use of domains has diffused to other parts of the world (see table 2.1). Although North American and European countries dominate this list, it also contains the top countries from every continent in the world with the exception of Africa. For most of the period from

Table 2.1 International distribution of domains, July 2003. Figures for China include Hong Kong.

Country	Country code	gTLD	ccTLD	Total	Per 1000 population	Percent of world's domains
USA	us	16,111,005	597,984	16,708,989	59.0	33.3
Germany	de	1,498,239	6,491,981	7,990,220	97.4	15.9
UK	uk	2,234,532	4,327,511	6,562,043	110.5	13.1
Canada	ca	1,070,259	352,800	1,423,059	46.2	2.8
China	cn/hk	982,665	318,181	1,300,846	1.0	2.6
Republic of Korea	kr	749,786	548,486	1,298,272	27.8	2.6
Italy	it	436,145	818,874	1,255,019	21.8	2.5
Netherlands	nl	333,224	904,011	1,237,235	77.8	2.5
Japan	jp	409,750	519,653	929,403	7.3	1.9
France	fr	735,497	168,538	904,035	15.3	1.8
Argentina	ar	51,189	750,000	801,189	21.7	1.6
Australia	au	353,500	347,576	701,076	36.7	1.4
Switzerland	ch	165,924	530,838	696,762	96.8	1.4
Brazil	br	79,118	488,295	567,413	3.3	1.1
Denmark	dk	54,822	438,863	493,685	93.2	1.0
Spain	es	402,291	48,933	451,224	11.3	0.9
Austria	at	90,313	287,194	377,507	46.6	0.8
Belgium	be	98,910	263,997	362,907	35.6	0.7
Sweden	se	187,467	172,953	360,420	41.0	0.7
Taiwan	tw	41,800	226,551	268,351	11.9	0.5
Total		28,129,402	22,121,541	50,250,943	8.4	100.0

Source: author's survey and country code domain registries; population figures are from 2000.

1994 to 2001 the USA also had the highest number of domains per capita of any of the top 20 countries. The USA only lost this ranking in January 2001 to the UK and later to Denmark, Switzerland, and Germany.

The variance in domain names per capita is quite marked, from a low in China of 1.0 per 1000 people to a high in the UK of 110.5 per 1000. While this reflects China's large population, part of these differences has to do with the country code domain registration policies in place in each country (OECD 1997). There are also significant variations in per-capita generic top-level domain (gTLD) name registrations between countries. Since gTLDs are all centrally registered

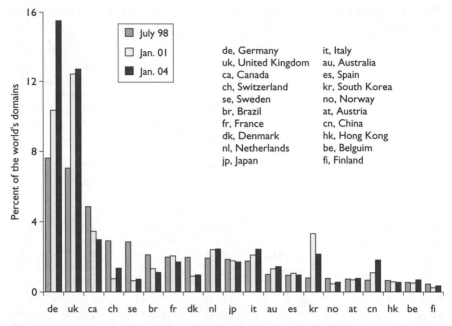

Figure 2.2 Countries' share of world domains (generic top-level domains and country code top-level domains), July 1998, January 2001, and January 2004.
Source: author's survey; order is based on July 1998 domain name totals.

under the same set rules, intercountry variations can point to significant differences between the Internet environments within countries.

Particularly notable is Japan's per-capita figure of 7.3 per 1000 people, which is the lowest in OECD countries and less than this sample's average of 14.7. Even limiting the analysis to just gTLD domains, Japan still has the lowest per-capita rate of any of the OECD countries. Although the exact cause of this relatively small number of domains is unknown, Kogawa Tetsuo, a professor of communications studies at Tokyo University of Economics, argues that Japan's strong tradition of centralized bureaucratic power is making Japan's adaptation to the Internet's amorphous structure difficult (cited in Rimmer and Morris-Suzuki, 1999). Aoyama (2001) further argues that Japan's unique history of consumer behavior has created a version of e-commerce structured around convenience stores rather than web sites.

It is also important to look at how these concentrations have changed over time. Figure 2.2 shows the percentage of the world's domains that the top 20 countries contained in July 1998, January 2001, and January 2004. To make this clearer, the USA, whose share changed from 49.2 percent in July 1998 to 42.3 percent in January 2001 and 32.3

Table 2.2 Percentage of the world's Internet domains in metropolitan areas ranked in terms of number of domains, July 1998 and January 2002.

| | | Percent of world's domains | | | | |
| | | July 1998 | | January 2002 | | |
City rank	Percent of world's population	Subtotal	Cumulative total	Subtotal	Cumulative total	Change in subtotal
Top 5	1.0	17.7	17.7	17.8	17.8	0.1
Top 10	1.5	6.8	24.5	5.7	23.5	−1.0
Top 50	4.2	21.7	46.2	19.0	42.6	−2.7
Top 100	6.0	9.1	55.3	8.1	50.7	−1.0
Top 500	11.7	13.6	68.9	14.9	65.6	1.4
Rest of world	88.3	31.1	100	34.4	100	3.3

Source: author's survey and country code domain registries; population figures are from 1996; US metropolitan areas defined by metropolitan statistical area/consolidated metropolitan statistical area; rankings are based on July 1998 data.

percent in January 2004, is not included. Although half of these countries experienced a relative drop in share from 1998 to 2004, many countries such as Germany, the UK, Italy, South Korea, and particularly China have significantly increased their share. Other countries, most notably smaller countries such as Switzerland, Sweden, and Denmark, have seen large relative drops. This likely reflects the already high per-capita rates of domain names in 1998 in many of these countries that made further expansion more difficult. In contrast, South Korea and China, as well as other countries not in the top 20 such as India, have significantly increased their role as suppliers of content and commerce to the Internet at the beginning of the 21st century.

Centrality of Urban Centers

Although country-level statistics give a good overview of a country's participation in the Internet, it is a very high level of aggregation. As table 2.1 suggests, countries with large populations, such as China, may mask significant concentrations of Internet activity within their major cities. Table 2.2 supports this contention by comparing the percentage of the world's population to the percentage of the world's

Map 2.4 Total number of domains (generic top-level domains and country code top-level domains) by city worldwide, January 2002.
Source: author survey.

Internet domains, both gTLDs and country code top-level domains (ccTLDs), in the top 500 cities in the world. Although the top 100 cities (45 of which are outside the USA) only contain 6 percent of the world's population, they contained over half of the world's Internet domains in January 2002.

Moreover, despite the 906 percent growth in domains during this time, the five largest regions in July 1998 (New York, Los Angeles, the San Francisco Bay, London, and Washington DC) exceeded this rate. The next 95 largest cities grew slower than the overall rate while the last two categories, cities 101–500 and the rest of the world, exceeded this rate. This suggests that while the registration of domain names has diffused outside its original concentration in the USA, the top metropolitan hubs have maintained their status.

Because of this continued dominance of certain cities and regions, map 2.4 shows gTLDs and ccTLDs located in major cities worldwide. Due to the relatively small size of the city database this map is biased toward larger cities. Although the distribution mirrors the location of major world cities, the size of London, at 1,182,928 domains, is particularly remarkable. The next largest cities are Seoul (560,796), Hong Kong (254,956), Berlin (233,303), Munich (229,736), and Paris (210,278). New York with 1,575,500 domains and Los Angeles with 1,463,900 domains are the largest concentrations of domains in the world and, with the exception of London (third), Seoul (sixth), Hong Kong (15th), Berlin (17th), and Munich (18th), the top 20 cities in the world in terms of total domain names are in the USA.

The global distribution of the Internet (in terms of both users and commercialization) at the beginning of the 21st century tells an important story of how a technology reputed to "render geography and cities meaningless" has developed a distinct geography with a contradictory pattern of wide dispersion of use and simultaneous concentration of its commercialization. While the Internet has increased the ability for isolated businesses or individuals to access (and be accessed by) the rest of the world, it also strengthens the ability of some companies to extend the scope and reach of their markets. In short, the Internet is not destroying geography but selectively connecting certain people and places into highly interactive networks, while at the same time largely bypassing others.

3

Mapping the Internet Industry

The selective connectivity of the Internet is based on its users, who remain rooted in specific social and geographic contexts and are shaped by the institutions, customs, and norms of these places.[1] This is particularly true for commercial endeavors, which are categorically more complicated than simply surfing the Web or emailing. Although anyone can create a simple web page offering goods and services, the ability to attract people to it is a fundamental challenge to any Internet-based business's success. Furthermore, the basic logistics of commercial activities, such as creating products, taking orders, or answering customer questions, present additional hurdles for would-be commercial web sites.

None of these barriers are insurmountable but their existence further belies the myth that Internet companies can flourish without concern toward location. Thus, despite expectations that firms would take advantage of the space-transcending ability of the Internet, the commercial Internet exhibits much of the traditional unevenness that has characterized urban and economic development throughout history.

Mapping Internet Industry Clusters

The commercial Internet industry at the end of the 20th century was largely based in the USA due in large part to the historical concentration of infrastructure and expertise that the ARPANET and follow-up projects engendered (Abbate, 1999). This concentration makes the USA a logical case study for further analysis of the clustering of companies engaged in the creation, organization, and distribution of Internet-mediated content, service, and commerce. A combination of three indicators – domain names, top web sites, and Internet firms – is used to identify the specific regional nodes of the industry.[2] The use of several independently developed datasets cross-checks the findings of any indicator against the others, providing an important degree of validation.

Domain names

The first indicator of the Internet industry is the distribution of domain names such as google.com or yahoo.com. Although the registration of domain names has diffused to other countries, they were predominantly located in the USA during the period from 1994 to 2000. As table 3.1 illustrates, there has been remarkable growth in the number of domain names as the Internet industry has developed. This marks the shift of Internet use away from large academic, military, or research institutions toward the commercial use of the Internet by businesses both large and small. As the commercialization process unfolded, the demand for domains, particularly those under the .com TLD, increased dramatically.

Maps 3.1–3.5 illustrate the diffusion of commercial domain names around the USA from January 1994 to July 2000. The earliest distribution in January 1994 demonstrates the legacy of the federally funded ARPANET and NSFNET programs, with the largest concentration of commercial domains located in Washington DC. At this point in time, the only people who even knew what a domain name was, let alone would wish to register one, were those who had contact with these projects. Other important ARPANET centers such as Boston, Los Angeles, and the San Francisco Bay region are also evident at this time, although they are smaller than Washington DC.

Map 3.2 shows the continued growth of domain names in the early centers as well as the emergence of new nodes. According to table 3.1, this period marks the fastest growth rate of domains as they expanded from 30,000 to 828,000. Particularly interesting is the overtaking of Washington DC by the Los Angeles, New York, and San Francisco metropolitan regions. These three regions have maintained this dominant status ever since.

The next three maps illustrate the continued growth and expansion of domain names. During this period, significant concentrations devel-

Table 3.1 Growth and US share of domain names.

Date	Number	Percent in USA
ccTLDs and gTLDs		
July 1992	16,000	–
January 1993	21,000	–
July 1993	26,000	–
January 1994	30,000	–
July 1994	46,000	–
January 1995	71,000	–
July 1995	120,000	–
January 1996	240,000	–
July 1996	488,000	–
January 1997	828,000	83.0
July 1997	1,301,000	–
January 1998	2,292,000	–
Just gTLDs		
July 1998*	2,154,634	74.8
January 1999	4,025,425	74.4
July 1999	7,052,350	69.3
January 2000	10,008,468	66.7
July 2000	18,648,629	59.7
January 2001	27,480,324	63.2
July 2001	30,089,731	60.2
January 2002	29,195,636	54.9
July 2002	27,131,204	55.7
January 2003	28,155,114	54.5
July 2003	29,968,266	53.6
January 2004	33,058,414	53.7

Source: July 1992 to January 1998 domain name totals (from Zakon, 1999) are for all types of domains, i.e., generic top-level domains (gTLD) and country code top-level domains (ccTLD); 1997 figures for gTLD domains in the USA from Quarterman (1997); all other figures from author's survey, which covers the com/net/org TLDs.
* The lower figure for July 1998 compared with January 1998 is because it only includes gTLDs.

oped in northern and southern California, the Northwest, the Eastern Seaboard, and scattered throughout the rest of the country. In many ways this pattern follows the distribution of population, with most cities emerging as notable sites of domain name registrations. But the distribution of the Internet industry is not simply a straightforward correlation to population. Closer examination at the metropolitan

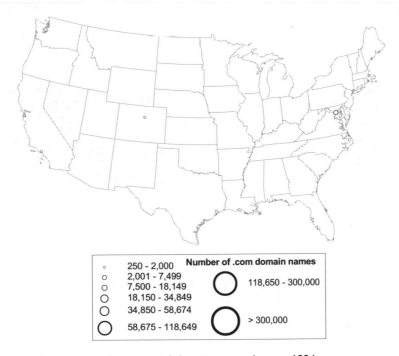

Map 3.1 Distribution of commercial domain names, January 1994.
Source: author survey.

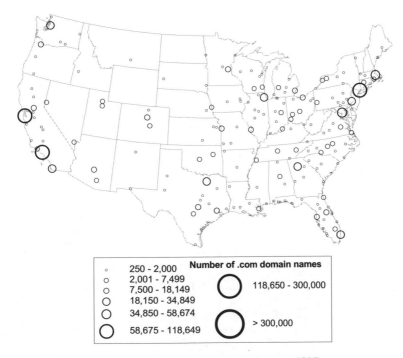

Map 3.2 Distribution of commercial domain names, January 1997.
Source: author survey.

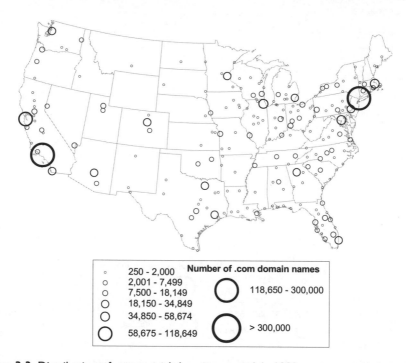

Map 3.3 Distribution of commercial domain names, July 1998.
Source: author survey.

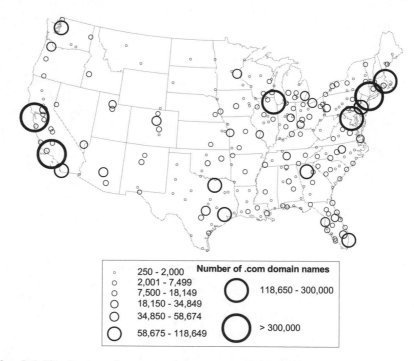

Map 3.4 Distribution of commercial domain names, July 1999.
Source: author survey.

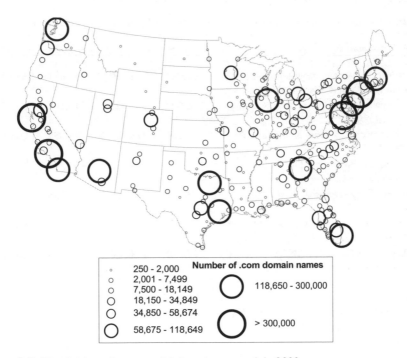

Map 3.5 Distribution of commercial domain names, July 2000.
Source: author survey.

statistical area (MSA) and consolidated metropolitan statistical area (CMSA) level reveals significant differences in the concentration of domain names. There is a noticeable discontinuity between the top three regions and the rest in terms of total numbers of domain names (see table 3.2). Together, the New York, San Francisco Bay, and Los Angeles regions have more gTLD domain names than the next 11 largest metropolitan regions combined.

Additionally, as table 3.3 illustrates, the intensity of the commercial use of the Internet varies considerably between regions. A very different ordering of metropolitan regions is obtained when the data are standardized by the number of domains per firm. The San Francisco Bay moves to the number one position and new regions, such as Provo, Utah, Austin, Texas, and Las Vegas, Nevada, appear as smaller but highly specialized areas of commercial domain names. These differences are very pronounced in comparisons among the largest regions. The San Francisco Bay area has more than twice the number

Table 3.2 Top 15 concentrations of domain names (gTLD) by CMSA/MSA, 1998–2001.

CMSA description	July 1998	July 1999	July 2000	July 2001
New York–northern New Jersey–Long Island, NY–NJ–CT–PA CMSA	168,066	552,750	1,241,871	1,645,875
Los Angeles–Riverside–Orange County, CA CMSA	146,697	553,325	1,116,142	1,535,325
San Francisco–Oakland–San Jose, CA CMSA	122,384	378,075	788,599	1,033,650
Washington–Baltimore, DC–MD–VA–WV CMSA	69,084	204,900	480,309	713,325
Chicago–Gary–Kenosha, IL–IN–WI CMSA	53,386	154,050	344,356	510,975
Boston–Worcester–Lawrence, MA–NH–ME–CT CMSA	55,411	158,850	332,502	484,125
Dallas–Fort Worth, TX CMSA	37,811	113,300	229,253	350,250
Miami–Fort Lauderdale, FL CMSA	38,009	122,400	242,108	348,300
Seattle–Tacoma–Bremerton, WA CMSA	35,999	100,125	226,061	333,675
Atlanta, GA	31,456	98,225	199,166	329,925
Philadelphia–Wilmington–Atlantic City, PA–NJ–DE–MD CMSA	37,768	105,850	229,323	323,250
San Diego, CA	31,244	130,150	185,365	293,700
Houston–Galveston–Brazoria, TX CMSA	28,390	87,475	180,063	261,225
Phoenix–Mesa, AZ	21,946	64,725	148,213	228,000
Denver–Boulder–Greeley, CO CMSA	25,415	67,050	142,933	219,675

Source: author's survey.

of domain names per firm as the Chicago, Philadelphia, Dallas, or Houston metropolitan regions.

Furthermore, as maps 3.6 and 3.7 demonstrate, domain names are not evenly distributed within regions but are clustered in particular locations. For example, there are high concentrations of domain names in the city of San Francisco, Manhattan, around San Jose, and in Silicon Valley. In addition, there are many smaller concentrations such as Berkeley–Emeryville in California and Park Slope in Brooklyn.

A final important difference between US regions is how their level of Internet activity has changed as the Internet industry developed. A useful technique for comparing regions over time uses a ratio that indicates the extent to which a region is specialized in domain names compared with the USA as a whole. A value greater than 1.0 indicates a higher number of domains per firm than the national average and a

Table 3.3 Top 15 specializations of domain names (gTLD) by CMSA/MSA, 1998–2001.

CMSA description	gTLDs per 1000 establishments			
	July 1998	July 1999	July 2000	July 2001
San Francisco–Oakland–San Jose, CA CMSA	406.1	1254.4	2616.5	3429.6
Provo–Orem, UT	349.2	1618.1	2103.4	3261.8
San Diego, CA	305.0	1270.5	1809.4	2867.0
Austin–San Marcos, TX	255.1	1002.3	1733.1	2813.4
Las Vegas, NV–AZ	233.8	835.3	1649.2	2752.0
Los Angeles–Riverside–Orange County, CA CMSA	243.8	919.4	1854.6	2551.1
Santa Barbara–Santa Maria–Lompoc, CA	233.4	699.7	1663.7	2274.5
Phoenix–Mesa, AZ	220.3	649.8	1487.9	2288.9
Washington–Baltimore, DC–MD–VA–WV CMSA	231.6	687.0	1610.4	2391.7
Seattle–Tacoma–Bremerton, WA CMSA	227.9	633.9	1431.3	2112.7
Miami–Fort Lauderdale, FL CMSA	219.7	707.4	1399.3	2013.0
Reno, NV	265.6	612.7	1140.4	2153.4
Boston–Worcester–Lawrence, MA–NH–ME–CT CMSA	230.2	659.9	1381.3	2011.2
Atlanta, GA	198.0	618.4	1253.9	2077.1
West Palm Beach–Boca Raton, FL	195.4	637.2	1375.3	2134.8
USA	149.6	454.0	1081.0	1592.0

Source: author's survey; establishment data is from Dun and Bradstreet, *Marketplace Data* CD, 1998.

value less than 1.0 indicates a lack of specialization.[3] Figure 3.1 provides an overview of how the specialization ratios of the top 10 CMSAs have changed since 1994. There has been a relative drop in specialization of San Francisco and Boston over time reflecting their central role in the precommercial Internet and the dot-com boom. Both regions retained a higher number of domains per firm than the national average, but the diffusion of knowledge about the Internet as a potential site for commercial use made the extremely high ratio in 1995 unsustainable.

Equally interesting is the relatively large increase in the specializations of Los Angeles and Miami. While it is impossible to provide a

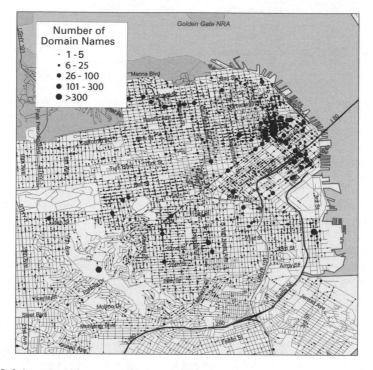

Map 3.6 Location of commercial domain names in downtown San Francisco, July 1998.
Source: author survey.

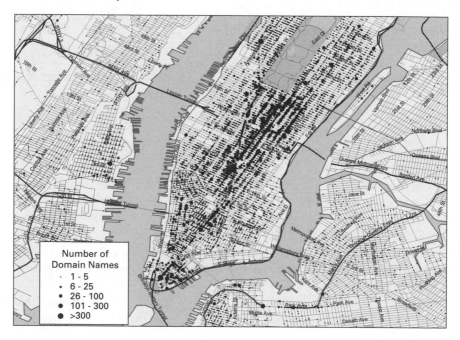

Map 3.7 Location of commercial domain names in downtown New York, July 1998.
Source: author survey.

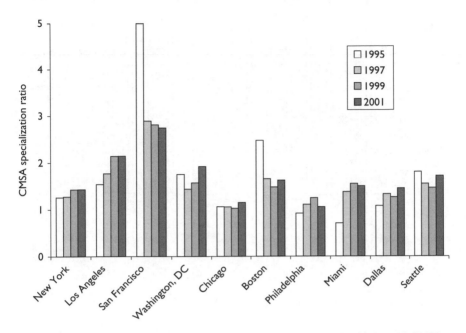

Figure 3.1 Change in commercial domain name specialization ratio for top 10 CMSAs, 1995–2001.

single causal factor for the increased specialization of these cities, these results illustrate the evolution of the Internet from its initial role as a technological development tool for the Department of Defense and academics to its application for broader commercial purposes. For example, many of most specialized CMSAs in 1994 such as Champaign–Urbana and Colorado Springs were closely associated with universities or the Department of Defense and quickly dropped in specialization as the Internet commercialized.

Top web sites

Lists of top web sites provide a second indicator of the Internet industry and an independent comparison to the analysis of domain name geography. First discussed by Paltridge (1997), rankings of top web sites are used to create a weighted measure of activity that provides a better indication of the most important web sites. A number of com-

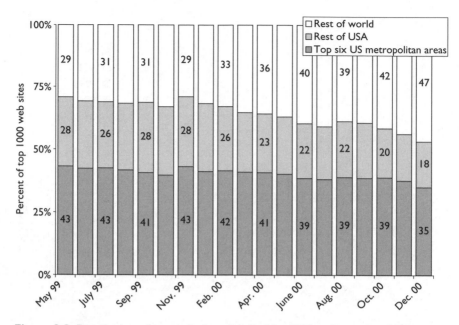

Figure 3.2 Distribution of top web sites globally, May 1999 to December 2000.
Source: based on Alexa Research web site rankings.

panies provide these rankings and this analysis relies upon the Alexa
Research ranking of the 1000 most visited web sites.

The distribution of these top web sites over time is illustrated in
figure 3.2. This period of 20 months also corresponds to the time
during which there was a significant increase in the number of non-
US Internet uses. The proportion of Internet users outside the USA
increased from 43 percent in July 1999 to 63 percent in July 2000.
During this same period the proportion of top web sites outside the
USA increased from 31 to 41 percent. These data suggest that the
growth of consumption of Internet information outside the USA is
occurring at a faster rate than the growth of the Internet industry
outside the USA.

Within the USA the distribution of top web sites mirrors the distribution of domain names presented in tables 3.2 and 3.3. The top three
regional concentrations of domains, New York, Los Angeles, and San
Francisco, are also the locations of the top 1000 web sites (see figure
3.3). In the case of top web sites, however, San Francisco leads the
urban hierarchy in terms of absolute numbers. Because the indicator
of top web sites reflects a weighted measure that is not present

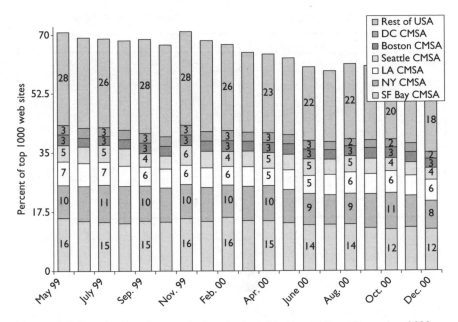

Figure 3.3 Distribution of top web sites in the USA, May 1999 to December 2000. *Source*: based on Alexa Research web site rankings.

in domain names, it strongly suggests that San Francisco is the leading region in the US Internet industry. Likewise, the location of top web sites shows that other regions such as Seattle, Boston, and Washington, DC, have a stronger Internet industry presence than cities such as Chicago, Philadelphia, or Dallas, despite a smaller overall number of domain names.

Finally, even at the subregional level the distribution of top web sites in the top two Internet regions, San Francisco and New York, is very similar to the pattern of domain names shown in maps 3.6 and 3.7. As map 3.8 illustrates, the same concentrations in South of Market in San Francisco, Silicon Valley, and downtown Manhattan shown by domain names are evident in the location of top web sites.

Internet industry firms

The final indicator used to corroborate the findings of the first two is a database of Internet firms from Hoover's Online. This database is

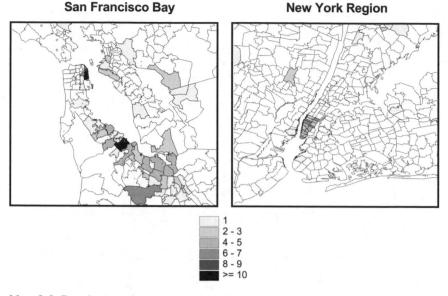

Map 3.8 Distribution of top web sites by zip code in the San Francisco Bay and New York regions, February 2000.
Source: based on Alexa Research web-site rankings; location determined by author.

useful because it is a direct measure of Internet industry firms as opposed to the indirect measures offered by domain names and lists of top web sites. Although it is important to acknowledge that this list is a selective sample rather than a complete population of Internet industry firms, it represents a systematic effort to compile a collection of the most active and important firms within the Internet industry in the USA.

As table 3.4 illustrates, the urban hierarchy identified by the first two indicators of the Internet industry is largely supported by the available data on Internet firms. With 63 percent of Internet companies located in the top six metropolitan areas, this supports the overall pattern of concentration of the Internet industry in a few locations and corroborates the urban hierarchy established through the use of top web sites. Additionally, the distribution of Internet firms at the subregional level in the top two Internet regions, San Francisco and New York (see map 3.9), is very similar to the pattern of domain names shown in maps 3.6 and 3.7 and top web sites in map 3.8.

Table 3.4 Distribution of Internet firms, May 2000.

	Dot-com firms	Employees
San Francisco Bay CMSA	137	30,536
New York Metro CMSA	101	24,637
Los Angeles Metro CMSA	46	6,585
Seattle CMSA	34	14,413
Washington DC CMSA	32	28,201
Boston CMSA	47	13,660
Other	231	61,869
Percent in top regions	63%	66%
Observations (*n*)	628	

Source: Hoover's Online database; location is based on location provided within the database.

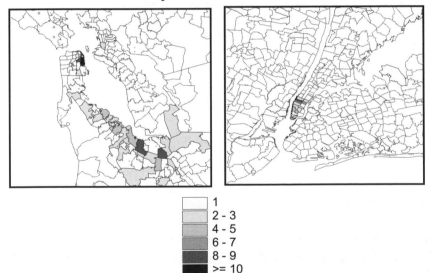

Map 3.9 Distribution of Internet firms by zip code in the San Francisco Bay and New York regions, May 2000.
Source: Hoover's Online database; location is based on location provided within the database.

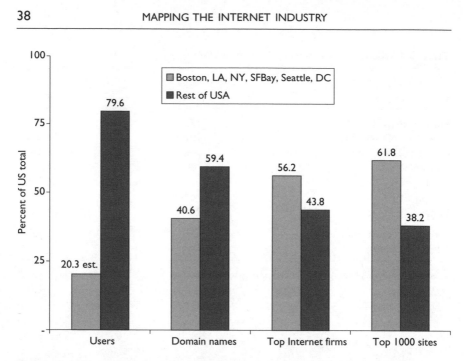

Figure 3.4 Concentration of Internet industry indicators in the USA, 2000.
Source: users from US census; domain names from author's survey, January 2000; top 1000 web sites from Alexa Research, February 2000 (geographic location is based on registration address of a web site's domain name); top Internet firms from Hoover's Online database, May 2000.

Many Regional Outcomes

The distribution of the Internet industry during the initial phase of commercialization, 1994 to 2000, shows a distinct pattern of clustering in certain urban agglomerations. Although the specifics differ depending upon which indicator is used, they all tell a very similar story. Figure 3.4 summarizes the geographic distribution of users, domain names, the top 1000 sites, and Internet firms between the top six Internet regions in the USA (Boston, Los Angeles, New York, San Francisco Bay, Seattle, and Washington, DC) and the rest of the country.

The first indicator, users, is often the one cited to demonstrate that the Internet is diffusing and it is clear that it is, with close to 80 percent of US web users located outside the major Internet centers. However, use of the Internet for email is a very different thing than the creation of dot-com companies. For example, the top six regions account for 40.6 percent of all domain names registered in the USA, contain 56.2 percent of US Internet firms, and house 61.8 percent of the top 1000 web sites located in the USA. Although this concentration of the commercial Internet is not unexpected given previous patterns of industrialization, the scale is quite remarkable.

Moreover, it clearly shows that the Internet industry is not simply bypassing geography but reorganizing the economic space in which businesses operate. This reorganization of economic space is reflected in the fortunes of specific regions, with some such as San Francisco Bay emerging as Internet industry hubs. While in some ways these top agglomerations correspond closely to existing city hierarchies, e.g., the global city of New York has developed into an Internet hub, there are significant discontinuities from previous patterns. For example, Philadelphia and Chicago are relatively small nodes and other urban agglomerations such as Austin and Las Vegas are emerging as new hubs.

However, these clusters do not simply appear by chance. The geography of any economic activity depends upon a number of factors, such as the nature of its production and inputs, and the Internet industry is no different. Information technologies do have the potential to disperse some activities, but the expectations of many people that the Internet would undermine the *rasion d'être* for economic agglomerations are unwarranted (Gilder and Peters, 1995; Negroponte, 1995, 1999).[4] While it is tempting to ascribe a single effect of the Internet on the economy, e.g., dispersal of jobs or industries, the experience of earlier technologies such as the telegraph or the electric engine (David, 1990) call for a more nuanced understanding of the Internet's interaction with the complicated geographies of production in capitalist economies (Storper and Walker, 1989).

4

Economic Clusters, Knowledge Circulation, and Venture Capital

The great paradox of the Internet industry is that despite the Internet's ability to transcend space, dot-com companies clustered in a relatively small number of urban agglomerations. Moreover, this pattern was in clear contrast to the rapid diffusion of Internet users worldwide. Although a number of factors contributed to this concentration, the difficulty with which knowledge about the industry was created and transferred served as a centripetal force and competitive advantage for certain regions. The Internet made standardized information and data widely available but this did not negate the value of face-to-face interaction, local norms, and local institutions in capturing and sharing nonubiquitous knowledge (Leamer and Storper, 2001).

Many regional development theories recognize the importance of knowledge and learning to economic development, and focus particularly on interfirm transactions as the mechanism for this knowledge generation. Considerably less attention is accorded to how financing and the actors associated with finance also contribute to knowledge creation and transfer. This is partially due to the fact that much of mainstream economic theory assumes the uniformity and perfect mobility of capital despite evidence that regional differences endure (Davis, 1966; Gertler, 1984; Clark and O'Conner, 1997). This often results in an implicit assumption that capital resources are fungible and interchangeable. These notions, however, are challenged by the characteristics of a particularly relevant form of financing for knowledge-intensive activities, i.e., venture capital. The clustering of venture capital resources in specific regions and the qualitative differences between venture firms (Bygrave and Timmons, 1992) demonstrates that financing can provide considerably more than simply money. Thus, while the Internet allowed information to be increasingly globalized, specialized knowledge (particularly that tied to financing) remained sticky and best transferred in spatial proximity.

Factors Behind Economic Clustering

The factors that underlie geographic clustering of firms and industries are numerous and include government-funded programs, supplies of skilled labor, the organization of labor, and innovative capacity. While these all played a role in the development of the Internet industry, they are insufficient in explaining the geography of these firms. For example, the predecessor to the Internet, the US military funded ARPANET, connected research universities that formed the initial nodes of this system. However, a deliberate "hands-off" policy by the federal government in the mid-1990s gave the private sector the lead role in commercializing the Web. This resulted in the eclipsing of key locations for the precommercial Internet, such as the University of Illinois in Champaign–Urbana, by other locales.

Likewise, the supply of labor has been used by many analyses to explain industrial location. Strong job growth in high-cost regions during the 1980s and 1990s is often attributed to factors such as the need for firms to remain close to suppliers and skilled labor (Schoenberger, 1988; Angel, 1989; Angel and Engstrom, 1995; Florida, 2002). This focus on skilled labor is quite relevant to the Internet industry. However, because the inputs required by the Internet industry were new, e.g., Java programming and selling advertising on web sites, there were no preexisting concentrations of labor with the needed skills and their emergence relied upon the adaptability of a region and its workers. Thus, while measures of human capital provide an indication of a region's potential to create Internet companies, it does not guarantee its ability to successfully do so.

A related explanation for industrial clustering is the availability and organization of appropriate labor skills in a region. For example, the theory of flexible specialization argues that short-term labor contracts and interfirm cooperation between small firms in certain regions provided companies with a competitive advantage in a rapidly changing economy (Piore and Sabel, 1984). While influential, flexible specialization theory has difficulty in explaining how a dynamic region of small firms differs from a region that contains small firms but lacks the dynamism of an industrial district. Again the existence of small firms and flexible labor arrangements does not mean that they will successfully organize to create Internet firms. One needs to consider how regions become sites for learning and knowledge creation that encourage experimentation and innovation.

Although long ignored by economists and geographers alike, innovation is the cornerstone of capitalist economic development that brings an inseparable combination of short-term instability and long-term growth (Schumpeter, 1939, 1942). While recognized as a basic "constancy of capitalism" (Storper and Walker, 1989), innovation is geographically "lumpy" as new ideas slowly permeate from person to person and firm to firm. Economic history is marked by numerous instances in which existing skills, production processes, industries, and locations were rendered obsolete by a new innovation. This disruptive nature of capitalist development is emblematic of the Internet industry in the late 1990s. As characterized in countless articles and attributed to any number of pundits the common creed was, "The Internet changes everything." Although this rhetoric was vastly overstated, the Internet did usher in an unprecedented amount of experimental and innovative behavior surrounding the commercial use of the Internet. The task is to explain why particular regions were leaders in this activity, thereby becoming the centers of the Internet industry.

Theories of Regional Institutions

Geographers have given a number of labels to regions that are successful centers of innovation, e.g., industrial districts, learning regions, entrepreneurial milieus, economic clusters, and milieus of innovation. All these theories share a focus on the way in which regional resources are created and organized to understand why some regions develop and others with a similar set of factor endowments do not (Castells and Hall, 1994; Saxenian, 1994; Porter, 1998; Malecki, 2000a). Researchers often focus on regions such as northern Italy, Denmark, Baden Württemberg, and Silicon Valley that contain clusters of small firms in horizontal relationships.

However, Saxenian (2000) argues that the presence of small firms or the concentration of interfirm transactions, such as those documented by Scott (1988a,b) or Acs and Audretsch (1993), cannot explain why a region is successful. To understand this, an analysis of the structure and nature of these interfirm relations is needed. In the case of Silicon Valley, which was consistently identified as a successful concentration of interfirm networks, Saxenian (1994) argues that these ties and the consequent blurring of firm boundaries that allow knowledge to spread quickly largely accounted for the region's dynamism compared to other places with similar resources. The approach advocated by Saxenian is often referred to as institutionalist (Amin, 1999), and functions through a detailed understanding of a region's social and professional institutions and norms of interaction to uncover the specific causal mechanisms that help create dynamic and innovative regions.

There is a wide range of approaches to studying institutions (Storper and Harrison, 1991; Storper and Scott, 1993; Saxenian, 1994; Locke, 1995; Herrigel, 1996; Storper, 1997; Aoki, 2000a,b; Kenney and Florida, 2000; Suchman, 2000), but all focus on local and nonmarket characteristics of regions, i.e., institutions, such as routines, culture, personal ties, cross-firm professional contacts, and tacit knowledge.[1] Regardless of how they are defined, these institutions form the rules and expectations in which firms and individuals interact and conduct business in a region. Through these interactions social networks are created that can foster environments in which knowledge is exchanged and learning is fostered. These social networks facilitate a number of interactions, ranging from simple information exchange to the creation of new companies. Regions with more or denser networks are better environments for all types of business activity. Of special interest for the Internet industry is the role of local institutions and social capital in creating and using knowledge about new technologies and business models.

Some, most notably Markusen (1999), have critiqued this approach as "fuzzy," with too much emphasis on processes rather than specific actors. While this line of critique injects a note of caution into institutional research, this approach can explain a great deal of the differentiation in regional development in the past 20 years. This is not to imply that more quantifiable factors do not play a role. Rather, it is argued that while factors such as governmental spending, skilled labor, and infrastructure are important foundations for regional development, they are not in themselves capable of explaining differences in regional growth. Instead, it is the process of combining these factors (and the knowledge of why and how to do so) that determines the growth potential and success of regional development. This is particularly relevant in the context of new and emerging industries such as the commercial Internet.

Tacit Knowledge in an Age of Global Information

Knowledge has long been recognized as an important factor in economic development, but often has been relegated to measures of education attainment or number of scientists per capita. It was Touraine's (1971) and Bell's (1973) work on the postindustrial society that first drew considerable attention to the importance of knowledge in an economy.[2] While Pred (1977) demonstrates that information and knowledge has played an important role throughout economic history, the growth in the power and ubiquity of telecommunications and computing technology during the past two decades casts a new spotlight on the use of knowledge in the economy (Castells, 1996). Contrary to expectations that the Internet would give people, regardless of their location, equal access to information, there has been increased attention to the issue of competitive knowledge. Maskell (2001) argues that as globalization made many inputs into ubiquities, i.e., costing the same for any firm no matter what its location, firms can best increase their competitiveness through their ability to manipulate and use knowledge.[3]

Efforts to bring knowledge into regional analysis often distinguish between codified and tacit knowledge based on Polanyi's (1958, 1967) work and his observation that "We know more than we can tell." Codified knowledge is defined as knowledge that is possible to record or transmit in symbols such as words, drawings, or other technical specifications or that is manifested in some type of concrete form such as a piece of machinery or equipment. In contrast, tacit knowledge is not easily captured in a transferable form, be it symbol or concrete, but is acquired through observation or interaction in which one largely learns by doing (Arrow, 1962). Because holders of tacit knowledge are not completely aware of what exactly they are doing and/or are unable to successfully express this knowledge symbolically, tacit knowledge is said to be "sticky" and is best transferred through direct experience (Von Hippel, 1994).

Lundvall and Johnson (1994) expand this simple dichotomy and outline a typology of knowledge that includes:

- know-what, a broad knowledge about facts which is very similar to information;

- know-why, an understanding of scientific principles;
- know-how, specific skills ranging from artisan aptitudes to the ability of business people to assess market opportunities; and
- know-who, the density and strength of social networks.

The first two types of knowledge fit largely into the explicit knowledge category of Polanyi's definition. The last two are more tacit in nature, although certain types of know-how can be captured in patents and copyrights. Nevertheless, know-how generally has some degree of context specificity and in many ways depends upon the interaction within and between firms and other local institutions such as universities for its creation. Lundvall and Johnson (1994, p. 30) argue, "One might say that important elements of tacit knowledge are collective rather than individual. Here takeovers and mergers may be regarded as attempts to gain access to tacit knowledge and know-how."

Their final category of knowledge, know-who, is explicit to the extent that information about individuals can be broadly distributed through news services and directories. However, having biographical and contact information for someone is very different from possessing tacit knowledge of a person built up through interaction. Moreover, the ability to readily access a person, particularly one who is quite busy, depends on the formation of some type of largely tacit social relationship. Although these types of knowledge are defined at the individual level, they can be used constructively at the firm or regional level.

The importance of context and the collective nature of know-how and know-who knowledge in innovative firm behavior has clear geographic implications and has been used by a number of researchers to explain the continued concentration of industries in particular regions. Although by definition every region contains a supply of tacit knowledge about its industries, people, and economy, this knowledge may be of little value to others, e.g., the recipes for food that the rest of the world finds unpalatable, or may not be organized in such a way that allows for its commercial exploitation, e.g., the long observed problem of transferring technology out of universities and research institutions. Therefore, regions in which valuable tacit knowledge is created and also successfully used gain important competitive advantages in a globalized economy.

However, Gertler (2003) argues that while tacit knowledge has increasingly drawn the attention of economic geographers, it is troubled by a relatively loose definition and the assumption that it has the uniform effect of geographically constraining business. The following section builds upon Gertler's work and explores the way in which tacit knowledge can be produced and transferred.

Producing tacit knowledge

Tacit knowledge is a difficult concept to measure and capture beyond simple metrics of educational attainment or patents, which have often been used to measure a region's knowledge and skill resources. Partly this is due to the fact that while individuals form the basic unit of knowledge, knowledge that can have an economic impact is created through social interaction. As Howells (2000, p. 54) argues, "The problem with harnessing knowledge . . . especially in relation to new and difficult knowledge associated with innovation, is precisely because it is so hard to share a common knowledge frame between different individuals . . . we all have different knowledge frames."

(a) Firms

This problem of harnessing the knowledge of individuals has generally been conceived as a firm-level problem and has caused some reformulation of economic theories of the firm. For example, in competence theory, firms are said to succeed based on their ability to create (via research and development) or acquire (via hiring people or connections) a unique stock of tacit knowledge to be exploited to their competitive advantage (Malecki, 2000b).[4] This makes the management of tacit information a key concern of firms and in the case of multi-location enterprises raises considerable challenges. Howells (2000, p. 54) argues that the differences "between knowledge frames are likely to be not all that great within a small, single-site firm staffed by people from a similar class and cultural background, but will widen as firm size . . . and geographic spread increases." This situation is at the heart of the knowledge management literature that has tackled questions of how firms can create knowledge, particularly in multilocation contexts (Teece, 1998).

At the same time, firms face the opposite problem of containing knowledge that is often in danger of spreading via the interaction and exchange of information through individual networks of practice. Classic examples of the leaky nature of firm boundaries include the migration of transistor technology from Bell Labs in New Jersey to

Shockley and Fairchild Semiconductor, the graphical user interface and mouse technology that left Xerox Park for Apple Computers, and the movement of the Internet browser from CERN labs to the University of Illinois to Netscape Communications. Although means exist (e.g., patents or copyrights) for firms to capture and control knowledge, it is a continuing containment problem since knowledge remains with individuals and often the most knowledgeable workers are also the most mobile. Moreover, precisely because tacit knowledge is difficult to codify, it is very hard to control and contain through intellectual property rights techniques. Many lawsuits have been filed precisely on the basis that an employee took proprietary knowledge. This was the reason Netscape completely rewrote the code for their browser, although it still ended up in court with the University of Illinois over the issue (Reid, 1997).

(b) Regions

Although firms are the preferred actor in neoclassical economic analysis, the ability of certain regions to foster knowledge creation has been an ongoing and parallel theme. Alfred Marshall (1890, p. 271) identified these concentrations over a century ago and argued that in addition to a supply of workers and subsidiary industries, these regions are advantageous for firms because, "The mysteries of the trade become no mysteries; but are as it were in the air, and children learn many of them unconsciously." Thus, economic geographers and regionalists view regions themselves as important actors in the knowledge creation process because they contain the firms, supply chain relationships, and networks of practice that can create new knowledge through interaction. All regions, however, do not contain similar resources and the ability to create new knowledge depends greatly on industrial and institutional structures built up over time (Storper, 1997).

In particular, the theory of flexible specialization postulates that innovation is stimulated as firms learn from one another and adopt best-practice technologies. Given the intensity and frequency of these interactions, physical proximity is seen as essential. This theory depends strongly on subcontracting and market transactions within regions embedded in larger systems of informal and nonmarket ties to account for the ability of firms to learn and adapt (Harrison, 1992; Saxenian, 1994; Gertler, 1995).[5] Nevertheless, these theories have largely focused on the role of interfirm producer–supplier transactions as a mechanism for knowledge creation and exchange, trust building, and innovation. This focus, however, neglects another aspect of Marshall's theories, i.e., the effect of variation.

Although Marshall's idea of "mysteries in the air" is well known, a related theory has received considerably less attention. Reintroduced into the discussion of regional economic development by Loasby (1999) and Maskell (2001), this theory highlights the variation in firm strategies and techniques that results from the fact that "shared understanding is always going to be filtered, perceived, stored and reconverted in our own individual 'knowledge frames' which are going to be slightly different" (Howells, 2000, p. 54). Marshall (1890, pp. 355–6) made a similar argument with the following observation.

> Even in the same place and the same trade no two persons pursuing the same aims will adopt exactly the same routes. The tendency to variation is a chief cause of progress; and the abler the undertakers in any trade the greater will this tendency be. . . . For instance, of two manufacturers in the same trade, one will perhaps have a larger wages bill and the other heavier charges on account of machinery; of two retail dealers one will have a larger capital locked up in stock and the other will spend more on advertisements and other means of building up the immaterial capital of a profitable trade connection. And in minor details the variations are numberless. Each man's actions are influenced by his special opportunities and resources, as well as by his temperament and his associations: but each, taking account of his own means, will push the investment of capital in his business in each several directions.

In contrast to theories based on relations and transactions between firms, this mechanism suggests a very different way in which agglomerations of firms create tacit knowledge. The potential variation in the implementation of tacit knowledge in a region by individual actors brings about a wider array of knowledge than would result if everyone followed the same method.[6] The result is similar to an experimental situation in which specifics of the final goals (i.e., marketable products) may be unclear and techniques to best achieve them (i.e., production process) are unknown, so simultaneous efforts are undertaken in multiple directions. The success and failures of these experiments are an important source of tacit knowledge for a region.

However, this is not an argument that producer–supplier and transaction-based relationships between firms are not an important source of knowledge creation in regions. The goal is simply to include the process of experimentation and observation in theories of how knowledge is created during the process of regional development. Firms within regions with dense concentrations of related firms will benefit from tacit knowledge gained through market and informal interactions with suppliers and in many cases competitors, but will also gain tacit knowledge through direct observation of the strategies and fortunes of its competitors without any explicit cooperation or relationship. These two regional knowledge creation processes are by no means mutually exclusive and regions in which firms are able to accrue tacit knowledge via both sources are likely to be the most dynamic and successful.

Transferring tacit knowledge

Although the creation of knowledge is a fundamental part of regional development, much of the research on tacit knowledge has focused on the difficulty of transferring it in market settings.[7] In part this is due to the fact that the value of tacit knowledge is difficult to predict before buyers have received it. While the seller has a good understanding of it, they cannot fully reveal it or the knowledge is transferred. This makes market exchanges of knowledge problematic except for some very codified knowledge manifested in patent rights, copyrights, or machinery. While this difficulty has led many researchers to emphasize the inherently local and nonmarket nature of tacit knowledge, Gertler (2003) cautions against this and argues that there are a number of ways in which tacit knowledge can be transferred. These include knowledge management within firms, communities of practice that transcend firm boundaries, and learning regions.

(a) Firms and communities of practice

While firms may be adept at creating tacit knowledge, albeit generally in one location, the transfer of this knowledge to other parts of the company can be extremely difficult. This challenge is readily acknowledged by large multisite firms and has resulted in numerous organizational strategies. For example, Nonaka and Takeuchi (1995, pp. 192–3) propose what they refer to as a hypertext organization that would "accumulate knowledge . . . by transforming knowledge dynamically between two structural layers – those of the business system, which is organized as a traditional hierarchy, and of the project team, which is organized as a typical task force." This quote suggests that this area remains one of considerable fluidity and experimentation and is echoed by many whose business is the commercialization of knowledge (Teece, 1998).

Brown and Duguid (2000) discuss how networks of practice, loosely connected people engaged in similar activities or with similar interests, can be the channel through which new ideas spread. They see these communities as existing alongside firm organization because the makeup of these groups, e.g., engineers, marketers, lawyers, transcend any one company. While Brown and Duguid are primarily concerned with single locations, these communities would not necessarily be limited to physically proximate groups (Bunnell and Coe, 2001; Gertler, 2003). Amin and Cohendet (1999, 2000) argue that relationships and occupational similarities are often more important for the transfer of tacit knowledge than physical closeness and Bunnell and Coe (2001) advance the idea that closeness in innovation is undergoing deterritorialization.

The extent to which knowledge can be transferred and its value is a matter of some debate. While arguably certain kinds of knowledge can be spread easily, e.g., know-what and know-why, it is not necessarily going to be the most valuable. Often networks of practice are most effective in transferring tacit knowledge when they operate in a local context because many types of know-how and know-who are based on the efficacy of an individual's personal relationships that are strengthened by face-to-face contact. Both Saxenian (1994), who documents the networks of practice among Silicon Valley semiconductor engineers, and Cohen and Fields (1999), who identify a number of networks such as universities, law firms, and business groups, emphasize the role of proximity in supporting these relationships.

Although telecommunications have greatly expanded the frequency, distance, and bandwidth of nonlocal relationships, this has not rendered local connections meaningless but largely increased the diversity of relationships in which anyone is involved. As Brown and Duguid (2000, p. 146) argue, "One of the most powerful uses of information technology seems to be to support people who do work together directly and to allow them to schedule efficient face-to-face meetings." Thus, while leaving open the possibility for nonproximate networks in which tacit knowledge is exchanged, current research shows that geographic proximity still plays a significant and even leading role in this process.

(b) Regions

Due to this fact, an important focus within economic geography has been upon the role of regions in transferring tacit knowledge. A number of different but intertwined strands of this research include learning regions (Florida, 1995; Maskell and Malmberg, 1999), institutionalists (Saxenian, 1994; Storper, 1997), and social capital theory (Putnam, 1994; Cohen and Fields, 1999). This research focuses on locally embedded and nonmarket characteristics of regions, such as culture, codes, and conventions, that create basic similarities between actors and allow for the building of trust and sharing of information. Because these networks transcend firm and organizational boundaries, the best way to take advantage of them is to be located within the region.

In her comparison between Silicon Valley and Route 128, Saxenian (1994) argues that it is precisely differences between regional institutions, namely attitudes toward sharing information, that explain each region's fortunes in the late 1980s and early 1990s. Maskell and Malmberg (1999, p. 181) echo her assessment and argue that, "It is the region's distinct institutional endowment that embeds knowledge and allows for knowledge creation which – through interaction with the available physical and human resources – constitutes its capabilities and enhances or abates the competitiveness of the firms in the region." This "learning through interaction" has increasingly been used to explain the existence of clusters of innovative industries and milieus of innovation.

However, the focus on interaction between firms, be they traded transactions or untraded interdependencies, has neglected the role of variation in regional economies. As Maskell (2001, pp. 9–10) argues, the ability of firms who are competitors or even in unrelated industries to observe one another's strategies and performance can provide a great deal of valuable information.

Co-localized firms undertaking similar activities find themselves in a situation where every difference in the solutions chosen, however small, can be observed and compared. While it might be easy for firms to blame the inadequate local factor market when confronted with the superior performance of competitors located far away, it is less so when the premium producer lies down the street. The sharing of common conditions, opportunities and threats make the strength and weaknesses of each individual firm apparent.

Firms do go to great lengths to safeguard strategic decisions and other proprietary information from their competitors. Thus, the norms, conventions, and networks emphasized by those looking at interfirm relationships also play an important role in determining the extent to which tacit information about this variation is spread. Maskell (2001, p. 10) goes on to assert that

> It is by watching, discussing, and comparing dissimilar solutions emerging from the everyday practices that firms . . . become engaged in the process of learning and continuous improvement, on which their survival depends . . . Promising avenues identified by one firm will soon be available to others and firms along the horizontal dimension of the cluster are constantly given the opportunity to imitate the proven or foreseeable success of others while adding some ideas of their own. In this process they are often significantly assisted by sharing a communal social culture including collective beliefs, values, conventions and language.

Thus, the variation that comes about from similar firms operating under different strategies offers a fundamentally different and yet complementary basis for knowledge creation in a region compared with interfirm transactions and dependencies. Maskell (2001) goes so far as to argue that the variation theory provides an explanation that at least in principle can explain clustering without any interfirm interaction. Just the ability to monitor and learn from other firms' behavior can provide a valuable source of tacit knowledge for companies. However, as Maskell is careful to point out, this theory need not be seen as a replacement for transactions between firms. Nevertheless, it provides a compelling mechanism for the transfer of tacit knowledge, and when combined with institutions and conventions of knowledge sharing and exchange can create significant momentum for regions in which it is located.

Venture Financing as Knowledge Circulation

A critique of the institutional and learning region approach is a tendency to focus on interfirm transactions to the exclusion of all other mechanisms for knowledge generation. Receiving considerably less attention is how the acquisition of financing also acts as a source of tacit knowledge creation and transfer. While acquiring capital is an issue for all types of new ventures, from restaurants to small-scale manufacturers (Friedman, 1995), it is particularly relevant to firms, such as those in the Internet industry, experimenting with innovations whether they be products, technologies, or business models. Most entrepreneurs begin their firms by bootstrapping, i.e., self-financing out of savings, but the combination of fast growth and high risk led many Internet companies to look for venture capital investors. Although venture capital investing represents a relatively small percentage of investment when compared with the total amount of business financing, it plays a crucial role in the process of innovation and the development of regional high technology clusters such as exhibited during the dot-com boom of the 1990s.[8]

This is aptly demonstrated by figure 4.1, which shows that the amount of venture capital invested in the USA increased by more than 1300 percent from 1995 to 2000. To put this in perspective, during the 18 months from July 1999 to December 2000, more venture capital was invested than in the previous 30 years.[9] The majority of this investment went into Internet-related companies, accounting for almost 90 percent of the increase between 1996 and 2000. In 1994, Internet companies were not even a distinct category from computer hardware or software; in 1995, the first year for which data on Internet-related investments are available, they totaled less than 20 percent of venture capital committed. By 2000, they accounted for 80 percent of venture capital investments.[10]

The mechanics of venture capital

While the bulk of entrepreneurs may prefer to self-finance their businesses or rely on bank and other types of debt financing in order to retain control of the company, this is not always possible or advisable. Zider (1998) argues that a combination of structural and regulatory aspects of capital markets makes it very difficult for young companies with viable ideas or technologies but without assets to gain access to the necessary capital to expand. Banks are constrained by usury laws and are unable to charge the level of interest for loans that the risk profile of these companies requires. Public markets and investment banks generally are not interested or cannot make investments in companies that have not reached a certain threshold of size, sales, and profits.[11] Such firms are good candidates for risk capital that fills a niche between an entrepreneur's ability to self-finance and the point at which banks and public markets can provide financing.

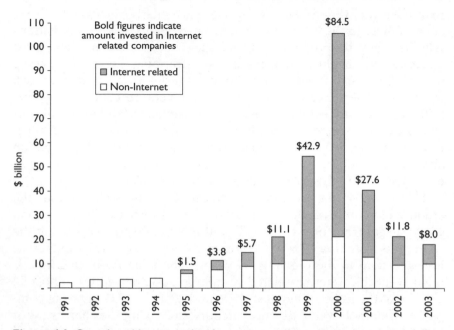

Figure 4.1 Growth and Internet-related venture capital investments in nominal dollars, 1991–2003. Bold figures indicate amount invested in Internet-related companies.
Source: 1991–1994, Venture Economics; 1995–2003, PwC/VE/NVCA Moneytree survey.

Risk capital has been available historically, but it was largely an ad-hoc system in which wealthy individuals backed entrepreneurs and firms that had come to their attention (Bygrave and Timmons, 1992; Kenney and Florida, 2000). While this system still exists today in the form of angel or individual investors (Mason and Harrison, 1994, 1999; Harrison and Mason, 1996), it has been accompanied by an increase in the sophistication and institutionalization of risk capital. Beginning shortly after the Second World War and greatly expanding during the 1980s, the informal process of risk investing evolved into more formalized venture financing institutions.

There are a number of stages at which firms seek capital and the source and the nature of the financing differs accordingly. At the seed stage, i.e., an exploratory investment to see whether an idea warrants further consideration, companies are primarily funded by individual entrepreneurs or angel investors who are often family or friends. It is not until the startup stage, i.e., product development, prototyping, and market research, or the first round, i.e., initial commercial production, that more institutionalized sources of venture capital begin to invest. This involvement continues through additional rounds of investing as a company expands production until it achieves some kind of liquidity breakthrough such as an IPO or acquisition by an existing company. At this point, when the risk is somewhat reduced, investment and commercial banks begin to participate in the financing of a company and initial investors may begin to sell equity in order to achieve profit.

Currently, the most widely implemented model of venture capital in the USA is the venture capital limited partnership that invests capital from a venture fund.[12] The capital for the fund comes from a number of limited partners, including pension funds, insurance companies, banks, foundations, endowments, corporations, and wealthy individuals, who usually provide little input on investment decisions. The general partners, i.e., the venture capitalists, are responsible for managing the fund and selecting companies for investment, and receive a yearly management fee that equals a few percentage points of the fund. Venture funds are established for a set period of time, most commonly 10 years, after which the capital gains are distributed to both types of partners, with the general partners receiving approximately 15–25 percent of the gains.

Because venture capitalists receive an equity stake in a company in return for their investment, they are ultimately concerned that their companies have some sort of liquidity event (i.e., IPO, sale to another company, etc.) within 7–10 years so that they can realize their gains. Firms that achieve significant growth can have an IPO and the venture capitalists will receive stock that can be sold in the public markets. Companies that are struggling and in which the founders have sold a controlling interest to their investors may have their senior management replaced to try to improve the prospects of the company. Firms that are failures generally have their assets sold, with venture capitalists receiving preferential treatment in compensation.

Due to the high rate of failure of early-stage companies, venture capitalists have developed a number of strategies to manage their risk. First, rather than making one or two larger investments, venture capitalists place smaller amounts of money in a portfolio of companies. Although the performance varies a great deal between venture capitalists, in general the expectation is that out of 10 firms, two or three will meet and/or exceed their projections, two or three will fail completely, and the rest will become what is referred to as the "living dead," i.e., companies that may be viable but are not growing at the hoped-for rate. By using a portfolio model, venture capitalists are able to achieve substantial rates of return despite high failure rates. Second, venture capitalists manage risk by coinvesting in companies with other venture firms and structuring their investments in a series of rounds. Third, companies are held accountable for meeting certain targets or metrics in order to receive additional money, thus allowing a venture capitalist to stop funding a company not making the expected progress.

The final and arguably most important mechanisms for managing risk are the evaluation process and venture capitalists' subsequent involvement with firms. Venture capitalists expend considerable resources on building their knowledge about companies, entrepreneurs, competitors, and market conditions before making investments and monitoring companies after investing. Gompers and Lerner (1999, p. 130) argue that "By intensively scrutinizing firms before providing capital and then monitoring them afterwards, venture capitalists can alleviate some of the information gaps and reduce capital constraints. Thus ... it is the nonmonetary aspects of venture capital that are critical to its success." This conclusion reflects a long-standing recognition of the importance of nonmarket factors to venture investing (Gorman and Sahlman, 1989; Bygrave and Timmons, 1992).

A common strategy used to leverage nonmarket factors, reduce costs, and increase the quality of information is restricting investments to nearby companies (Gompers and Lerner, 1999, p. 180). By investing locally, venture capitalists can better use their networks of contacts to source and evaluate deals, provide timely assistance to their portfolio companies, and most importantly access tacit knowledge that would otherwise not be available. In a survey of 464 venture capital-funded entrepreneurs from 1967 to 1982, Timmons and Bygrave (1986, p. 161) found that "capital" was consistently the least important variable in an entrepreneur's decision. Entrepreneurs were primarily interested in establishing connections with venture capitalists who could help recruit management, provide contacts with other key actors, lend credibility to their company, and contribute to overall firm strategy. Although not exclusively limited to local relationships, this type of hands-on assistance is greatly facilitated by proximity.

Regional economic development and venture capital

The implications for regional development of this local orientation spurred considerable work on documenting the location, spatial mis-

Table 4.1 Distribution of venture capital investments, 1995–2000.

Region	Share of investments (%)	Share of dollars invested (%)
San Francisco, CA CMSA	28.8	32.3
Boston, MA CMSA	8.9	8.2
New York, NY CMSA	8.8	9.5
Los Angeles, CA CMSA	5.9	5.8
Washington, DC CMSA	4.6	5.2
San Diego, CA MSA	3.1	2.6
Seattle, WA CMSA	3.1	2.6
Denver, CO CMSA	2.8	4.1
Atlanta, GA MSA	2.7	2.0
Philadelphia, PA CMSA	2.5	2.1
Chicago, IL CMSA	2.1	2.0
Dallas, TX CMSA	2.0	2.3
Austin, TX MSA	2.0	1.9
Minneapolis, MN MSA	1.6	1.2
Raleigh–Durham, NC MSA	1.5	1.0
Total	80.4	82.8

Source: PwC/VE/NVCA Moneytree survey; aggregated to MSA/CMSAs by author.

match, and flows of venture capital investing in both the USA and elsewhere (Leinbach and Amrhein, 1987; Florida and Kenney, 1988a; Green and McNaughton, 1989; Martin, 1989; Green, 1991). As table 4.1 illustrates, the concentration of venture capital activity is significant, with the top 15 metropolitan areas accounting for around 80 percent of all venture capital investments between 1995 and 2000. The principal center of venture capital investment, the San Francisco Bay region, continues to receive the largest proportion of investments, although its share has dropped from the levels reported by Florida and Smith (1990) during the 1980s.

In addition to documenting this spatial concentration, researchers have incorporated venture capital within existing regional development and innovation theory (Florida and Kenney, 1988b; Thompson, 1989; Malecki, 1990) and modeled (often with mixed results) the relationship between the location of venture capital firms and the location of venture capital investment (Green and McNaughton, 1989).

Particularly influential is Florida and Kenney's (1988c) theorization of venture capital as a third way in Schumpeter's dichotomy of corporate versus individual entrepreneurism. They argue that venture capitalists act as catalysts or "technological gatekeepers" who facilitate and direct innovation in regions with strong social structures of innovation, i.e., concentrations of human capital, universities, and public research and development (Florida and Kenney, 1988a,c; Kenney and Florida, 2000). The importance of localized networks is supported by other research within both the geographic and business literatures (Bygrave, 1988; Malecki, 1990; Bygrave and Timmons, 1992; Gupta and Sapienza, 1992; Elango, Fried, Hisrich, and Polonchek, 1995; Friedman, 1995) which argues that the use of local networks is crucial for the exchange of specialized knowledge as well as for the direct involvement of venture capitalists in their portfolio companies (Gorman and Sahlman, 1989; Sapienza, 1992; Gompers and Lerner, 1999).

Thus, while capital in the most general sense of the word, i.e., money, provided the fuel for many Internet companies, in many ways it was the transmission of the tacit knowledge of the venture capitalist that was perceived as the more valuable element. In the end, it is precisely this variation in regions' venture capital systems to create and transmit tacit knowledge that is central to understanding the geography of the Internet industry.

5

Connecting Venture Capital to the Geography of the Internet Industry

During the dot-com boom the acquisition of venture capital was viewed as a strategic asset for both its monetary and nonmonetary inputs that could provide quick and competitive boosts to companies. Entrepreneurs sought to both capture the advantage of speed offered by venture capital and gain access to their local networks and knowledge. This influx of capital into venture funds had a major impact on the level and extent of entrepreneurial activity in the regions (such as the San Francisco Bay) in which it was concentrated. Responding to the wildly successful public offerings of early Internet companies such as Netscape and Yahoo! in 1995 and 1996 (both of which were funded by risk capital), venture capitalists jumped on the opportunity of the Internet, and began to fund and be approached by a wide variety of dot-com entrepreneurs.

However, this activity was highly place specific and regions with existing venture capital systems and histories of technological industries had a great advantage. Thus, despite telecommunications and global financial markets that expand the geographic range of economic interaction, regions and the dynamics of local capital markets are central in understanding the geography of the Internet industry.

Entrepreneurial Incentives to Get Big Fast

A key lesson taken from the first Internet companies such as Netscape and Yahoo! was the great advantage that accrued to first movers. Thus, acting quickly and managing time was perceived as one of the greatest challenges facing entrepreneurs. As one former entrepreneur turned angel investor noted in 1999:

> The Net has changed everything. You don't have to have great products. You can have mediocre products, it's really about marketing and partnerships. That is the most important thing today, getting as many people to know and have emotional equity with your company as possible, getting as many people to feel religious about your company's success, to take it personally. That's it! It's getting people emotional and getting people to have a vested interest in your success.

Although Internet entrepreneurs relied upon other regional resources, such as skilled labor and management recruiters, it was the influx of capital into these firms that facilitated the expansion of companies. Traditional time horizons of 5–7 years from startup to functioning company were dramatically compressed and cornerstones of company evaluation such as profitability and price of stock to earning ratios were supplanted by a pursuit of market share and "eyeballs," i.e., visitors to a web site. As Freeman (1999) argues in his analysis of Silicon Valley, "The issue here is speed. It is time. It's almost to the point that it matters less what you do than when you do it. An important part of the venture capitalist's job is to move this along rapidly, to make the right decision at the right time."

Looking for local and "smart" capital

Venture capitalists with established capital funds, existing networks, and experience in starting and growing companies were perceived by entrepreneurs as bringing an important competitive advantage to the quest to "get big fast." Continued spectacular performances by other venture-backed Internet firms, such as eBay and Amazon, which got very big very quickly, brought growing interest from investors who poured increasingly large amounts of capital into venture funds, and from entrepreneurs, who drafted tens of thousands of business plans.

This increase in activity only made the premium placed on time more important and further emphasized the advantages that came from getting capital that was also well connected. Given the concentration of venture capital resources in a few regions and the local orientations of many venture capitalists, the emphasis on venture financing strongly contributed to the concentration of Internet firms. In fact, the increased emphasis on time reinforced the tendency of venture capitalists to compress time through space and made them less interested in distant investments. As one San Francisco-based venture capitalist joked, "Why should I even drive down to Redwood City [approximately 25 miles south of San Francisco] to see a company when I have more quality business plans than I can hope to review here?"

However, securing the venture capital to finance the rapid growth of a company often did not happen as quickly as founders hoped. Many entrepreneurs faced the choice of going further in debt or shutting down operations completely. One San Francisco-based Internet company founder simply calls the process, "the bad spiral of hunting for money. Not only are you going longer without income but you are spending more and more money as time goes by." At issue is a problem of finding people (other than the founders) who see the potential business as a worthwhile, albeit risky, investment, and can validate the company. This is what one founder colorfully refers to as, "breaking through the bullshit barrier between the blow-hards and the doers"; it was seen by many entrepreneurs as the principal challenge facing their companies early in development. This founder of a Manhattan-based Internet service company argues, "If you don't get that seal of approval, then you're done. It's a pretty clear thing." Many other founders reflected this sentiment, including one based in San Francisco, who notes "Getting people, angels, venture firms, anyone to take us seriously . . . that by far was the hardest thing we did."

Thus, a common strategy, argues one Redwood City entrepreneur, is "to make sure that when one pot of money turns you down that you are able to find another. A lot of people either come to the Valley or spend a tremendous amount of time out here because this is . . . what did that guy say about robbing banks? . . . This is where the money is." As a result, regions with more risk capital available or in which it was easier for entrepreneurs to get access to this capital (not necessarily the same thing) were better environments for the entrepreneurial activity surrounding the creation of the Internet industry.

Particularly important was the speed with which capital could be acquired. As a San Francisco entrepreneur argues:

> The one reason and one reason only that there are so many companies out here is because this is where the capital is. It allows you to move fast, which is key since Internet time is seven times as fast as any other kind. Capital attracts companies, companies attract like companies and people, but they only attract them because there is capital here. It all revolves around capital.

While company founders listed a number of other challenges, such as recruiting management and other skilled workers, creating a marketing plan, and courting customers, these issues were often perceived as closely tied to a company's financial situation. As the founder of a Silicon Valley business-to-business company remarks:

> I'm saying that relative to the challenge of how to find the people to fuel the company, the technical challenges, the sales challenges were not as great. If you have enough sales people you can make enough sales, if you have enough engineers you can build stuff. If you have no money, you can't have enough engineers. It's hard, especially in a game where time is everything . . . and time is everything in the Internet space. The sooner you get funded, the faster you can hire resources, the faster you can get a solution to market and the faster you can create distance between you and the next company, which is what the race is all about.

In addition to obtaining it, entrepreneurs were also concerned about the source of their capital and distinguished between "smart money" and "dumb money." Smart money comes from people, generally venture capitalists or well-connected angel investors, who have an expertise in a particular sector or technology and have connections and networks to other companies who are potential customers, suppliers, or partners. In addition to providing a company with money, which is the only contribution of dumb money, smart money can help companies in any number of ways. As the founder of a San Francisco e-commerce firm argues:

> Smart money is always the only money you want. And what does that mean by smart money? It means that the person has a massive Rolodex. That's really what it means and they may not know squat about your business but if they can get doors opened for you at Netscape, Eudora, or Lotus or Microsoft, they are worth their weight in gold.

However, getting the most from a venture capitalist or "smart money" is constrained by geography because many venture capitalists prefer to invest locally or in partnership with another venture capitalist that is near the firm (Florida and Kenney, 1988c; Florida and Smith, 1993; Saxenian, 1994). The interaction between geography and venture capital funding was well recognized by the entrepreneurs interviewed. The founder of an e-commerce company based in the San

Francisco Bay is certain that his location played an important role in his ability to secure capital.

> You can't be anywhere. To start companies you need to raise capital and investors would prefer to make investments locally because they have to spend time with the companies. I know some venture firms that say, "If I can't drive there within an hour, I don't make the investment." Especially in an early stage company, you want to have regular contact with the company, so access to capital drives a lot of decisions. Investors prefer to invest locally because they're always the ones on the plane having to travel to a company.

Other managers of Internet companies also highlight their location as a strategic choice. The CEO of an Internet software company reports that her venture capitalists told her, " 'You have tremendous value just by being in the Bay area.' We have better access to the venture community, a high quality venture community which makes the partnerships easier." Many venture capitalists also cite accessibility to capital as important to the future of Internet firms.

> Access to capital is a strategic weapon. Just look at a company like Amazon that just raised a billion dollars in debt. The ability for a company to fundraise fast, and then recruit and assemble a team fast is an advantage. I just think that part of how the venture capitalists help is that they are all just lined up on the same corridor and it's easier. People literally can meet someone at a moment's notice and when you're trying to get an hour's worth of a venture capitalist's time, which is pretty precious today, you're just more likely to meet with a venture capitalist just because it is more convenient for you to drive down 280 than for you to hop on an airplane to come out here.

This local orientation is borne out by a simple correlation between the number of venture capital offices and the number of venture capital investments at a range of geographic levels. As table 5.1 illustrates, there is a statistically significant correlation between these two variables at all levels of geography, from 5-digit zip codes to MSAs, and it increases as the geographic scale expands. Moreover, this geography is even stronger for earlier stage investments. By concentrating on nearby investments, particularly critical at early stages, venture capitalists are able to work more closely with companies and take advantage of local networks of contacts to lower cost, gain tacit knowledge, and manage risk.

Table 5.1 Correlation between venture capital (VC) offices (1999) and investments (1999 to third quarter 2000).

	All VC investments		Early-stage VC investments	
	Correlation	Number of observations	Correlation	Number of observations
5-digit zip code	0.298*	1920	0.286*	1316
4-digit zip code	0.502*	1057	0.541*	782
3-digit zip code	0.748*	387	0.808*	387
MSA	0.773*	184	0.817*	184

Source: number of VC offices from *Pratt's Guide to Venture Capital*, 2000; VC investments from PricewaterhouseCoopers Moneytree survey.
* significant at the 0.01 level.

The persuasiveness of capital agglomeration

This recognition that locating near sources of venture capital provides better access to the funding, networks, and advice of venture capitalists caused many Internet companies to view their location as a competitive advantage. Although most of the founders started their companies in the same place they had been prior to becoming an entrepreneur, many argued that knowing that there was ready access to capital made it much easier to take the risk of starting a company. One described the San Francisco Bay region as "a caldron of financing" which "enticed you to take a chance." In addition to the effect that local venture capital had within its region, many entrepreneurs noted that it also served as an attraction for people to relocate near it.

> I speak to some CEOs in Austin and Atlanta and Chicago and you hear about the fact that there is this growing venture community in Austin and to a certain extent, Atlanta, and they say flat out that they're reconsidering location because they're afraid they're not going to get the next round of financing. Just the concentration of money in Silicon Valley can be persuasive.

This "persuasiveness" is confirmed by entrepreneurs who chose to relocate to the San Francisco Bay area either prior to starting or after founding an Internet company elsewhere. While this became particularly intense during the commercialization of the Internet, it is a process that has long been bringing people to regions like the Bay area.[1] The founder of an Internet services company in San Francisco relocated from the east coast at the start of the 1990s because it seemed to be a place where he could explore some of the business plans he

> kept in a crazy idea folder. I grew up in New York, worked there, and I quickly learned that it [his job at a top-tier investment bank] bored me. I always wanted to start a company and I wanted to work with a younger company. I think it was the idea of California that made me want to move out here . . . Out in San Francisco the entrepreneurs are the rock stars and the whole system revolves around them. It's all set up to plug money into your crazy ideas.

This accessibility to capital and the means to explore new ideas also proved highly influential on the decisions of entrepreneurs who relocated to the San Francisco Bay to start companies after the Internet boom had begun. Although any location decision is based on a number of factors, including personal preferences, business connections, labor supply, etc., entrepreneurs consistently cited the availability of capital as a leading variable in their decisions. As the cofounder of a San Francisco-based Internet software company remembers, the list of possible locations was relatively short in his mind.

> When we looked where there was capital there were really four or five areas. The three big ones in order were, San Francisco, Boston, New York. . . . The second wave had Austin, Atlanta and Seattle. We really only saw six areas and three really big ones where starting a company from nothing and growing it was really possible. Those three we felt were relatively equal in having great talent pools, but San Francisco had much better access to capital.

Another company founder who moved his firm of half a dozen people from Toronto to Palo Alto echoes this sentiment. Largely this was because he felt that in order to succeed he needed to be in the center of venture capital that was interested in investing in Internet companies.

> The difference between there and here is black and white. In Toronto when I would meet with VCs, I would spend a lot of time trying to explain why the Internet was so important, trying to educate them as to what an opportunity it was. Often I spent so much time doing this

that I never even got to present my business plan. They didn't get it. So we came out here to get close to the venture capital that knew something. If you're an aspiring actor you go to Hollywood and if you're an Internet company you come to Silicon Valley. Out here they just get it and you can spend your meeting actually going over your business plan.

The power of the agglomeration of venture capital in the San Francisco Bay area is particularly striking when compared with other regions, since it is by far the largest concentration of venture activity in the country. As a New York-based entrepreneur notes, "Getting money meant going to California because no one in New York would talk to you. You'd talk to the VCs and they'd tell you that they didn't know that this [the Internet industry] existed. They couldn't believe that it was a phenomenon." This difficulty in gaining access to investors is particularly ironic since New York is a global center for finance.[2]

Eventually, the fortunes of these companies increased supplies of risk capital available around the country. Still, for many entrepreneurs it was difficult to get access to smart capital from an investor who was equipped to play a lead role in a company's strategy. The founder of a New York Internet services company interviewed in 1999 is quite adamant on this point.

Will raising money be easy this time? I don't know. Yes, there are more funds available. People say it's easy to raise money right now. You hear it over and over. You read it over and over. Just put a business plan together and line them up and pick your favorite, but it ain't true. Here's what's true. Once you get a lead investor it's easy. But I've got ten candidates for lead investors, all of them say, "I like it but we just don't have the bandwidth for this thing." So, I've got plenty of people who are willing to put in a million and a half when we get the lead investor. So in that sense maybe there is more money available but I still need the lead and there are still only a handful of guys who can be leads. The new names can't really be leads and the ones that can are more busy than ever.

This issue of "bandwidth," i.e., the ability for venture capitalists to spend time working with a company rather than just pumping money into it, is one of the limiting factors for the venture-backed model (Freeman, 1999). It is also why regional venture capital systems develop slowly and why simply pumping in additional capital into a region will not necessarily produce the dynamism of established venture capital centers.

Testing the Role of Venture Capital

To test the findings from interviews, a multivariate linear regression is used to explore the explanatory power of a number of regional attributes in the distribution of the Internet industry. However, the quality of the data used in this analysis prevents it from proving the relationship between the activity of venture capitalists and the location of the Internet industry. Although the most important input from venture capitalists are their networks, connections, and ability to work with companies, it is not possible to obtain a reliable measure of this and instead this analysis relies upon the number of venture capital investments in a region as a proxy.[3] In addition, this simple measure of size masks a great deal of differentiation in regional venture capital systems in terms of sector, stage, and involvement. Therefore, this analysis is best seen as an effort to reject the findings of the interviews that argue that venture capital played a leading role in the location and creation of these firms. The inability of these regressions to do so strengthens the argument that venture capital investing played an important role in determining the location of the Internet industry.

Introducing the variables

This analysis is conducted at the regional level, defined as either MSAs or CMSAs where available. Because of data availability issues with the dependent variables, the models contain approximately 90 regions where any venture capital investing has taken place. The goal of this analysis is to match factors of labor, educational attainment, and venture capital investments in existence in 1998 (midway through the commercialization process of the Internet) to outcomes in the year 2000.

The analysis uses two dependent variables for the location of the Internet industry (see table 5.2). The first dependent variable is the *number of top 1000 web sites* (based on the Alexa Research survey) located within a region in February 2000. A second variable, which was developed independently from the first and is more representative of dot-coms, is also considered. This second dependent variable

Table 5.2 Summary of variables.

Variable	Description	Source	Mean	Standard deviation	Minimum	Maximum
Dependent variables						
Top1000 web sites (log)	February 2000	Author	0.87	1.06	0.00	5.08
Top Internet firms (log)	May 2000	Author	0.82	1.11	0.00	4.93
Independent variables						
Total employment (log)	Size of region/external economies, 1995	US Census*	6.03	1.13	4.05	9.11
Number of patents per job	Ability to create commercially viable knowledge, 1995	US Patent and Trademark Office*	0.45	0.33	0.07	2.30
Population with BA/BS	Availability of skilled labor, 1990	US Census	0.15	0.04	0.08	0.30
Percentage of jobs in high-technology industries	Connection between the Internet industry and high technology, 1995	US Census*	0.03	0.02	0.002	0.12
Percentage of jobs in informational industries	Connection between the Internet industry and information process, 1995	US Census*	0.08	0.03	0.03	0.23
Domain name specialization ratio	Early mover advantage, 1994	Author	0.67	0.84	0.00	4.49
All venture capital investments (log)	Size of venture capital activity, 1997–98	PwC	2.47	1.33	0.69	5.76

* US Census data estimated for 1995.
PwC, PricewaterhouseCoopers.

is based on the database of *Internet companies* developed from Hoover's Online and only includes companies that were founded explicitly to use the Internet in their business.

The independent variables (see table 5.2) were selected to represent regional factors that have long been identified as supporting regional economic development. The first one, *total employment*, is simply a measure of size of the region and an indicator of a region's external economies. The second variable, the *number of patents per employee*, measures a region's ability to support the creation and commercialization of new knowledge. Finally, this analysis includes the *percentage of the population with a BA/BS degree* as a measure of the supply of skilled labor. In addition to these three variables that are supportive of knowledge-based development in general, the models include a number of specifically Internet-related variables.

Given the reliance of the Internet industry on computer technology, the analysis includes the *percentage of a region's jobs in high-technology industries*.[4] A related sectoral argument is that the Internet industry is less connected to high-technology jobs as traditionally defined, and more involved with information processing jobs that fall across many different industrial sectors. This is represented by the *percentage of a region's jobs in informational industries*.[5] Because these two labor force variables share some sectors in common, they are not included in the same models but compared with one another.

Some regions in the USA had an early introduction to the Internet through involvement with ARPANET and/or NSFNET in the 1980s and early 1990s. Simply being one of these early centers could provide a region with a head start that would provide its Internet industry with an advantage in developing quickly. This factor is represented in the *commercial domain name specialization ratio in 1994*. This ratio is similar to a location quotient and measures the extent to which a region was specialized in the use of the Internet before the commercialization process started.

The final independent variable is the availability of venture capital in a region, measured by the *total number of venture capital investments in 1997 and 1998*. This time period is used to reflect the time lag between venture capital investment and the performance of a company. Historically, venture capitalists expected that it might take up to 7 or 10 years for a return on their investment through some kind of liquidity event. During the commercialization of the Internet, however, this time horizon shrank and companies went from initial investment to IPO in as little as 2–3 years.

Because the two dependent variables and the independent variables of total employment and venture capital investments are highly concentrated in a few regions, the natural log for these variables is used to create a more normal distribution. Additionally, two outliers in terms of the number of venture capital investments, even when using natural logs (the San Francisco Bay and Boston), are excluded from the analysis in order to create a more linear model.[6]

Findings

This analysis uses multivariate linear regression to examine the relationships between the variables and understand how the two indicators of the Internet industry relate to the various measures of a region's environment. Each of the dependent variables was regressed against a number of combinations of the independent variables.[7] The results of these models are outlined in tables 5.3 and 5.4. In general, these regressions support the idea that venture capital investments and early involvement in the Internet are important factors in determining the geography of the Internet industry. The findings are less clear-cut on the role of existing high-technology or informational industries and educational levels. It found no significant relationship between the patents and the dependent variables. The models generally all had adjusted r^2 values above 0.50, suggesting a robust relationship between the variables.

Five different combinations of the independent variables regressed against the first dependent variable are outlined in table 5.3. Overall, the findings are quite robust, with adjusted r^2 values above 0.50 for all models. As expected, the measure of a region's size is positively correlated with the number of top web sites and is statistically significant in a majority of the models. The most consistent finding in these models is for venture capital investments. Although many of the permutations of the model considered are not shown in table 5.3, the coefficient for venture capital investments is consistently positive and significant at the 95 percent confidence interval and higher. This significance remains constant from simple models that only include total employment for a region, to more complex regressions involving several other indicators of a region's labor force, knowledge, and history.[8] These results suggest that venture capital investment in a region during 1997 and 1998 is positively and significantly correlated with the number of top web sites located in the region at the beginning of 2000.

Table 5.3 Regression findings: top 1000 web sites.

	Model 1		Model 2		Model 3		Model 4		Model 5	
	B	t-value	B	t-value	B	t-value	B	t-value	B	t-value
Dependent variable: log of top web sites										
Independent variables										
Log of employment	0.63	8.55***	0.59	7.73***	0.12	1.09	0.52	7.28***	0.18	1.63
Log of number of all VC investments, 1997–98					0.53	5.26***			0.42	3.91***
Location quotient of .com domains, 1994							0.47	4.33***	0.30	2.73**
Percentage of population with BA/BS	0.63	0.31	0.81	0.39	0.17	0.63	0.53	0.28	0.42	0.89
Number of patents per jobs	(0.09)	(0.29)	0.29	1.09	0.09	0.38	0.06	0.22	(0.02)	(0.07)
Percentage of high-tech jobs	9.23	2.26*								
Percentage of informational jobs			5.39	1.83	2.60	0.98	0.97	0.34	0.39	0.15
r^2	0.51		0.50		0.62		0.59		0.65	
Adjusted r^2	0.49		0.48		0.60		0.56		0.63	
Number of observations	92		92		92		92		92	
F-value	22.7		21.9		28.3		24.8		26.6	

*, significant at the 0.05 level; **, significant at the 0.01 level; ***, significant at the 0.001 level.

A second clear finding, although slightly less consistent than the results for venture capital investing, is historical involvement with the Internet. The indicator of a region's domain name specialization ratio in 1994 is consistently positive and in most models, simple or complex, statistically significant. This suggests that early centers of the Internet were at an advantage over other regions in producing web sites that were the most visited in 2000.

The results for the percentage of a region's jobs in high-technology industries are positive and are significant in the first model. This suggests a positive correlation between centers of high technology and successful Internet firms. However, if models 2–5 used the variable measuring the size of high-technology industry rather than infor-mational industry, the significance of high-technology industries disappears although its coefficients remain positive. The size of a region's informational industry remains positive in all the models but does not appear to be statistically significant with this dependent variable.

The variables of educational level and proprietary knowledge within a region are not significant in explaining the distribution of top web sites. Although the coefficient for educational level remains positive in all the models presented here, it does not emerge as a significant variable. This is somewhat surprising given that many researchers have found that educational levels correlate with increased entrepreneurial activity (Florida, 2002). One possible reason for this is the relative age of this variable to the others, particularly the depen-dent variables, and changes in regional educational levels since the decennial census of 1990 could be basis for this. Also there is some correlation between this measure of educational levels and the other variables. While not debilitating to this analysis, they do point to the limits of this dataset and could also account for this finding.

The results for the second dependent variable, regressed against the same independent variables, demonstrate much of the same relation-ships noted in the first set of models. The same five combinations of independent variables are outlined in table 5.4. These models tend to be even more robust than those of the first dependent variable and generally have adjusted r^2 values that are higher than those found in table 5.3. The most consistent finding is again the correlation between venture capital investing in a region during 1997 and 1998 and the number of Internet firms located in it by mid-2000.[9] The coefficient for this variable is consistently positive and is significant at higher levels than the regressions with the first dependent variable. Likewise, an

Table 5.4 Regression findings: Internet companies.

	Model 1		Model 2		Model 3		Model 4		Model 5	
	B	t-value	B	t-value	B	t-value	B	t-value	B	t-value
Dependent variable: log of Internet companies										
Independent variables										
Log of employment	0.67	9.39***	0.63	8.82***	0.09	0.99	0.55	8.55***	0.15	1.61
Log of number of all VC investments, 1997–98					0.61	6.97***			0.50	5.47***
Location quotient of com domains, 1994							0.49	4.96***	0.29	3.07**
Percentage of population with BA/BS	2.51	2.06*	1.84	0.94	0.43	0.27	0.43	0.25	0.84	0.55
Number of patents per jobs	0.08	0.28	0.33	1.28	0.09	0.45	0.07	0.32	0.01	0.05
Percentage of high-tech jobs	5.12	1.29								
Percentage of informational jobs			6.94	2.51*	3.75	1.65	2.32	0.88	1.63	0.72
r^2	0.56		0.58		0.73		0.68		0.76	
Adjusted r^2	0.54		0.57		0.72		0.66		0.74	
Number of observations	92		92		92		92		92	
F-value	28.3		30.9		47.8		36.3		45.3	

*, significant at the 0.05 level; **, significant at the 0.01 level; ***, significant at the 0.001 level.

early history of Internet involvement is positively and significantly correlated with a region being the location of Internet firms in 2000.

Interestingly, the results for the other variables measuring the quality of regions' labor force, proprietary knowledge, and involvement in the high-technology industry are a bit different with this dependent variable than the first. Whereas the percentage of the population with a bachelor's degree was never significant when regressed against the number of top web sites in a region, it is significant in the first model with this dependent variable. Additionally, the percentage of high-technology employment in the region has not emerged as a significant variable in terms of the location of the top Internet companies. The size of a region's informational industry, however, is positively and statistically significant in the second model. Although its significance drops when variables for historical involvement with the Internet and venture capital investing are included, this suggests that the two dependent variables diverge in some interesting ways.

Discussion

The findings of these regressions support the idea that venture capital has played an important role in the development of the Internet industry. In addition to the most basic level of access to money that the variables in these models measure, venture capital has contributed to the clustering of the Internet industry by its provision of a number of nonmonetary inputs such as management advice, contacts, and mentorship. In many ways these are what entrepreneurs value most about receiving venture capital (Timmons and Bygrave, 1986). The ability of venture capital to supply this type of value-added input quickly is dependent upon the quality of its networks. The role of spatial proximity in the diffusion of information and construction of social networks is particularly important in understanding this type of regional development and remains true even in a global economy.

It is also suggests that participation in the Internet during its precommercial phase provides regions with an advantage over others in the creation of successful Internet firms. As Abbate (1999) and Townsend (2001a) document, the Internet and particularly its predecessor ARPANET was originally concentrated in a few US Department of Defense-funded computer science departments in major research universities. These regions contained concentrations of people who were among the few to be aware of the Internet and its commercial potential. One result is that the creation of the World Wide Web's "killer app," the Mosaic browser which introduced graphical capabilities, took place in the relatively small town of Champaign–Urbana, Illinois, which also happened to be one of ARPANET's original nodes. Of course, this head start did not guarantee that a region would continue to be a major node in the commercial Internet. In the case of Mosaic, the entire team of its original developers were moved en masse to Silicon Valley to form the nucleus of Netscape Communications, which was instrumental in inspiring much of the commercializing efforts (Reid, 1997; Clark and Edwards, 1999).

There are also interesting differences between these two sets of models in the significance of high-technology employment and the educational level of a region. While high-technology employment is positively and significantly correlated with the number of top web sites in a region, employment in informational industries is positively correlated with the number of Internet firms in a region. Although the two dependent variables are related, and in fact strongly correlated, these findings demonstrate some important variation between these indicators. While the variable of top web sites does include firms that focus exclusively on Internet content production, they also include the web sites of companies that are popular with many of the Internet's users. Since the Internet has long been the domain of computer aficionados, is not surprising that many of these popular sites include older high-technology companies such Intel, Apple, and IBM. This suggests that the correlation between high-technology employment and top web sites may be more indicative of the popularity of high-technology web sites than a clear causal relationship between high technology and Internet companies. This is supported by the lack of significance for this variable in the second set of models that uses a more select definition of the Internet industry.

The correlation between the number of Internet firms and the percentage of the population with BS/BA degrees supports an observation often made concerning dot-com companies. Although they are based on the use of technology, many of these companies are not technology companies per se. Rather they leverage the technology of the Internet to reinvent or restructure existing business. Thus, rather than just needing a supply of highly skilled engineers or programmers, their labor needs include a much broader set of skills and hence the stronger and more positive correlation to general educational measures.

More than Money

This analysis shows that the clustering pattern of the Internet industry is indeed closely tied to venture capital investing. However, this should not be taken as a pure supply-side argument in which simple access to capital equates with entrepreneurial success. Rather venture capitalists' intricate connection to a region's knowledge, labor, and industries are what allowed it to play a central role in producing key firms in the emerging Internet industry. As Martin (1999, p. 11) argues, capital brings with it a whole set of social relations that color its value and use.

> [M]oney is not just an economic entity, a store of value, a means of exchange or even a "commodity" traded and speculated in for its own sake; it is also a *social relation*. Financial markets are themselves structured networks of social relations, interactions and dependencies – they are communities of actors and agents with shared interests, values and rules of behavior, trust, cooperation and competition. Face-to-face contact, personal recommendations and informal word-of-mouth have always been central to the conduct of financial business and transactions, and remain so even in an age of advanced telecommunications – geography matters. The social relations are an important part of the embedded micro-regulation (accepted mores, norms, customs and rules) of business practice and behavior in financial institutions and markets.

This emphasis on money as a social relation captures venture capitalists' use of systems of personal contacts and networks to exchange scarce information, assess business plans, and back startups in a quick and efficient manner. Far from being an easily moved and fungible commodity, venture capital investing depends upon nonmonetary inputs such as knowledge and investors prefer to be close to companies in order to monitor and assist them. While the commercialization of the Internet would have no doubt taken place without the efforts of venture capitalists, it is likely that it would have been much slower and would have had a significantly different structure (Mandel, 2000).

6

Finance and the Brokering of Knowledge

The connection between venture capitalists and the expansion of the Internet industry makes understanding the operation of venture capital essential. Venture capitalists spend a great deal of their time building and reinforcing their social networks (know-who) in order to monitor the activity of other venture capitalists, firms, and markets. This allows them to offer a number of nonmonetary inputs (know-how), such as management advice, contacts, and mentorship, in addition to the money they invest; and in many ways it is the nonmonetary inputs that entrepreneurs value most (Timmons and Bygrave, 1986).

Thus, venture capitalists can be characterized as knowledge brokers who acquire and create intelligence about industries, market conditions, entrepreneurs, and companies through a constant process of Marshallian interaction and observation. This knowledge is then used to select promising industries, find good firms, and assist portfolio companies.[1] The Internet industry provides a good example of the value of knowledge, because venture capitalists in the San Francisco Bay region were among the first to identify the Internet's commercial potential and quickly spread this lesson throughout the region. While people in other places saw the potential of the commercial Internet, particularly after Netscape went public in 1995, this knowledge did not have the same impact because the long history and system of venture capital present in the Bay area did not exist in other regions.

Selecting Promising Industries

In the back of every venture capitalist's mind is the search for the next big technological change that will alter a significant portion of the economy, provide super-profits to the firms they support, and elevate their reputation to the likes of Arthur Rock, Don Valentine, John Doerr, and Vinod Khosla. This focus on industries, however, contrasts with what Zider (1998) describes as the myth about venture capital in which any good idea can receive funding if it has a passionate and skilled entrepreneur behind it. Zider (1998, p. 131) argues:

> The reality is that they [venture capitalists] invest in good industries –
> that is, industries that are more competitively forgiving than the market
> as a whole. In effect, venture capitalists focus on the middle part of the
> classic industry S-curve. They avoid both the early stages, when tech-
> nologies are uncertain and market needs are unknown, and the later
> stages, when competitive shakeouts and consolidations are inevitable
> and growth rates slow dramatically.

The search for these promising industries is based on a combination of know-how on emerging technologies and business plans, connections to people (know-who) equipped to evaluate risk and benefits, and direct observation of the variation in companies funded by other investors.

This results in venture capitalists attempting to both stay in step with other venture capitalists to leverage their experience, but at the same time be one step beyond the investing horizon in order to capture the largest potential gains. During the expansion of the Internet industry, venture capitalists went through several investment foci, such as business to consumer, business to business, and fiberoptics and telecommunications. As one Menlo Park venture capitalist explains, "VCs tend to work in packs. They are well networked and know when and what people are looking at, so by the time the press release comes out, most people already know that something was in the works and some of them even know the details and already may be looking to find a similar company." This exchange of knowledge about investing patterns largely depends on what a New York venture capitalist refers to as "personal connections with a number of people but nothing formalized between VC firms." Although there are serious efforts to keep investments secret until the agreements are finalized, there is a great deal of rumor and speculation. Moreover, once a public announcement is made or leaked, this knowledge spreads very quickly through the venture capital network.

The impact of this on entrepreneurs and startups can be quite intense. Many founders of Internet firms recount the difficulty and importance of getting their first "real" venture capital investor to validate their company. As the CEO of a Manhattan Internet services company recalls:

> You know, venture capital in every level is sort of a game where if one person likes you, they all like you. If one person hates you, they all hate you. It's a funny game. When Dawntreader said, "We like you guys and think that your technology and plan makes sense" all of a sudden, and I mean instantly, the other venture capital guys in town were, "I want in on this deal, too!" and at that point we were able to pick and choose.

This shift in venture capitalists' thinking, and sudden increases and drops of interest, is echoed by the founder of a Redwood City business-to-business Internet company who had a tough time selling his idea in 1997: "Later on in the year, peoples' understanding of e-commerce changed to using the Internet to solve business problems rather than retail problems and suddenly our phone calls were getting returned." While news of a venture capitalist's investments quickly spreads outside of a region, local investors have better access to the tacit knowledge about the company and its industry. This information is acquired largely through what one investor characterizes as "individual contacts in other firms and you get together as regularly as you can. You talk about what's happening, what's working, what's not. That's your pipeline for information and relationships for building syndicates."

A major problem of following the herd is catching the tail end of an investing trend and is why a venture capitalist's personal stock of tacit knowledge is so valuable. Just as firms' tacit knowledge loses its competitive value as it becomes more widely codified and distributed (Maskell and Malmberg, 1999), the tacit knowledge of venture capitalists, particularly about a hot technology or business strategy, loses its value as word spreads. This makes local and quick access to knowledge a very important part of venture capital investing. As a Los Angeles venture capitalist argues:

> There are top tier VC firms who really work at understanding a technology and potential markets and really set the pace for the rest of us. The real problem about following is that if there are already 250 optical system companies then you are already too late. So you have to do your own research because you don't make money on just following the herd. You want to be near the herd so you can see where it's going but you don't want to be in the center just following along and you really don't

want to be at the end. You need to think these things in terms of five to seven years but it's hard to think that far out and you want to know what the others are doing. That's why we always race to hear the rumors about what the other VCs are doing.

This dilemma of following the herd and yet striking one's own path is a principal challenge facing venture capitalists and success in doing this is what builds reputation and provides the ability to raise subsequent venture funds.

Although the significant drops in venture capital investing post 2000 reflect the downturn in public markets, the venture capital investing process did not stop but focused on the search for the next new industry, technology, or business strategy. Regions that contain greater variation or experimentation within new industries, albeit some of which will be spectacularly unsuccessful, contain more tacit knowledge about a greater number of different models all of which can be easily observed. As Von Burg and Kenney (2000, p. 1152) note:

> The difference between a radical innovation with massive capital gains and a mistake with no chance of success is not always easy to discern a priori. Many apparently sure things and great entrepreneurial visions ultimately look foolish, because they find no customers, encounter problems that cannot be solved technically, or come to fruition only years or even decades after the first investments.

This point is particularly relevant for the Internet industry because of the great uncertainty surrounding the viability of business plans and the number of spectacular failures that occurred in 2000 and 2001.

Using Tacit Knowledge to Find and Select Firms

Having identified promising industries, venture capitalists must decide in which individual companies to invest. Gompers and Lerner (1999, pp. 127–8) characterize the effort to overcome uncertainty about the future prospects of firms as the main task of venture capital. This is particularly challenging because venture capitalists must make decisions in a context of asymmetric information where entrepreneurs are much more knowledgeable about a firm's operations and prospects. As a result, venture capitalists expend considerable energy attempting to close this gap.

However, largely missing from the business literature on venture capital investing is the role of proximity in providing a means of accessing information, a standard tactic.[2] For example, a Palo Alto venture capitalist notes that when it comes to investing in companies located outside the region, he considers it essential to have some reliable source of knowledge about the firm.

> We find companies all over the place and one of our considerations is whether we know people there and if not, we have to fly there, and generally we have not made investments outside of Silicon Valley. We've considered it, San Diego or Irvine, but it's a day out of your time and it would be a bit of a learning curve to break new ground in another place.

While this investor recognizes the possibility of gaining knowledge by a nonspatially based connection, he wants to have that person be familiar with the firm's local environment or feels compelled to travel and pursue the information himself.

Moreover, limiting investments to nearby firms provides easier and faster access to an entrepreneur's references, which can often be double-checked by a venture capitalist's own personal connections and knowledge about similar businesses under consideration by competitors. As a San Francisco venture capitalist notes:

> I just don't bother looking outside, really. To a large extent if I had my preference, I would stay local just because the business is a resource constraining business. I view venture capitalists as factories and the more efficient you become at looking at deals and doing the due diligence the better off you are. When you go travel to the east coast to look at a company it just takes two days out of your week and I don't think the background is as good.

This investor argues that since he is more efficient at using his local connections, he can get better information faster upon which to base his decisions. The following section explores how tacit knowledge and local connections are central to venture capitalists' ability to first generate a sizable deal flow, control and screen this flow, and ultimately make informed investment decisions.

Generating deal flow

One of the basic tasks for venture capitalists is maintaining a steady stream of entrepreneurs and companies looking for funding. This is referred to as "deal flow" and reflects both the awareness within the entrepreneurial community of sources of venture funds and the active efforts of individual venture capitalists. Although a certain amount of deal flow comes simply from setting up shop as a venture capitalist, much of what comes unsolicited is of questionable quality and little interest. Venture capitalists want to see the deals with the most promise and depend heavily on their connections and relationships in the larger entrepreneurial community to find them. As a Palo Alto venture capitalist notes, "I'd say primarily the way you get deal flow is that you build your own network of people who look for you. What comes over the transom, through the website, from emails is generally not very qualified."

This emphasis on relationships, i.e., the know-who of individual venture capitalists, includes the possibility of deals generated via long-distance connections built up over a venture capitalist's career. Although venture capitalists report hearing about deals located elsewhere that were of potential interest, the majority saw local connections as more capable of generating quality deal flow because these ties were stronger and more trustworthy. A relatively new venture capitalist in New York argues that local connections were central to his ability to get deal flow.

> We have relationships here that told us flat out that you would not get this deal if you were doing this from afar. It's being able to initially make the contact and then establish the relationship . . . and ultimately socialize with the person. You can't do that over the phone. To really succeed you still need to have a presence. All this on-line stuff, yeah makes us easily assessable, but it's not the same as being able to go out and talk across the table.

The key know-who held by venture capitalists, i.e., the personal and idiosyncratic relationships of each individual, determines the level of "traction" that venture capitalists have in a region's deal flow. "Traction" is a term used to denote the strength and extent of a venture capitalist's connections and is often tied to a relatively long career history within a particular location. As a Menlo Park investor argues:

> As a venture capitalist one of the basic things you need is traction. A lot of it is getting in the Valley deal flow and getting in front of the best stuff and getting yourself out there. I think I'm just a person who has been in the Valley forever, and am well networked into the Valley, went

to Stanford, worked at Apple and knew a lot of these companies and I do a lot of networking. I spend a lot of time out there and try to make myself part of the fabric of the community.

Having a history of venture investing within a region provides another important avenue for generating deal flow, i.e., the entrepreneurs that the venture capitalist funded previously. Due to their active involvement in a company, technology, and/or market sector, these entrepreneurs often see new companies that are potential competitors, customers, or suppliers before a venture capitalist does. As one Palo Alto venture capitalist explains, "We have a group of people who are in interesting positions like VP of business relations at eBay and president of AboveNet and others like that who see every single Internet company that's interesting and that's coming to market, just by being in their position, and they just swing deals our way when they think it's interesting for us." While a venture capitalist's know-who can extend well beyond the local region, local contacts usually remain the most productive in producing deal flow.

Screening deal flow

In addition to using their know-who to generate deal flow, venture capitalists rely upon these relationships to prioritize the high volume of business plans submitted to them. This volume can be quite large, particularly for venture capitalists and firms who have established reputations. Because they already depend on their relationships to see the most promising deals, venture capitalists use their knowledge about the source of the deal to judge whether a business plan is worth examining. As a Menlo Park venture capitalist explains:

> I'm currently getting maybe four or five business plans a day and each business plan if you're really going to look at it, is a ten hour commitment and you need some way to make coherent decisions on what to look at. I'm already getting the opportunity to do 40 or 50 hours of work per day. So I've got to have some way to triage that and I depend on someone I know to alert me to good deals. If I don't know this person at all and they're coming in totally cold, they have to say something really compelling to get me to look at it. I mean there has to be something like "Hi I'm the founder of USA Networks and I'm starting a new company" or something like that. But frankly those people aren't going

to cold call me either. A lot of what I call the random deal flow is coming from people who aren't here, and just don't understand the Valley. They just hear that we have a big pot of money out here, so they send you some not-ready-for-this-kind-of-venture-capital deal.

The reliance on personal connections makes this screening technique both highly effective and provides the first cut in the company evaluation process.

This mechanism is well known among entrepreneurs, who depend either on their own reputation and connections to venture capital to arrange meetings with venture capitalists or look for someone else to make the initial introduction.[3] This introduction is by no means a guarantee of funding, merely a way for a company to get a venture capitalist to review their idea and business plan. As the founder of a San Francisco based e-commerce company remembers, "I was initially upset that I needed my brother to make the introduction but I realized there is just no getting through the door without an introduction, and the introduction will only get someone's attention for a brief moment. The rest is up to you."

In many ways, this screening process is not limited by proximity as a venture capitalist located in New York may look at a company based on a recommendation from someone they respect regardless of their location. In practice, however, connections used to make these types of introductions are often locally based. As the founder of an Internet software company in Manhattan notes, "You really have to have an in somewhere and because we're in New York, we had a lot of connections through our work and personal lives within the Wall Street community and it just made sense to look there." Thus, even in a largely relationship-driven part of a venture capitalist's work, proximity is relevant.

Evaluating and selecting deals

The final step before venture capitalists make an investment is the evaluation and selection process. These decisions rely heavily upon local tacit knowledge to research the background of an entrepreneur and company but also involve observing the investments made by other venture capitalists. This process can involve multiple meetings between the founders of a company and any number of venture capitalists to go over the business plan in detail. In the case of very strong

business models or founders' reputations, the evaluation and investment can take place very quickly. In contrast, venture capitalists spend
a significant amount of time doing due diligence for startups coming
from less illustrious backgrounds while simultaneously monitoring
what other and potentially competing venture capitalists think about
the prospects of the company.

As the cofounder of a Foster City Internet software company
remarks:

> What we started to notice is that you do these presentations, and you
> get to a certain point, a certain level of interest and then you kind of get
> stuck. You really need someone to pull it together and be the catalyst
> for bringing other VCs and getting you funded. Until that happens, until
> there is a spark, it's difficult to get things moving.

The decision to make a spark or pull the trigger on the part of venture
capitalists largely depends upon tacit knowledge about the managers
of the company. Even when venture capitalists invest in distant companies, they generally do so through coinvestment with other venture
capitalist firms located near to the company who act as the lead
investor. As Florida and Smith (1993, p. 448) argue, "Co-investment
facilitates capital flows and, in doing so, loosens the spatial constraint
on investment." While this is true in the sense that capital flows from
major financial centers such New York and Chicago to regions like the
San Francisco Bay area, the spatial constraint remains in that these
flows are most often channeled through a venture capitalist on the
ground.

The easiest way for a venture capitalist to make a decision that is
the least dependent on tacit knowledge is based on the well-known
reputation of the founders of a company. As one Menlo Park venture
capitalist remarks, "I am a firm believer that history repeats itself and
those who have been successful before will be successful again." Individuals who have founded companies and are interested in doing
so again are commonly referred to as "serial entrepreneurs" and are
highly valued by the investment community because they have
proven that they have the necessary skills to succeed. This emphasis
is almost entirely on the reputation of the entrepreneur rather than an
evaluation of the business per se. This was the case for the first commercial Internet company, Netscape, whose founder Jim Clark was
able to use his previous success with Silicon Graphics to quickly line
up venture capital for his browser company and later secure funding

for two other Internet companies, Healtheon and myCFO (Clark and Edwards, 1999; Lewis, 2000).

But this type of reputation and track record is not available for the majority of the companies that pass the initial screening process. These are firms with a plausible business plan, some management in place and even product lines, but do not have a track record in startup companies. The founder of an Internet commerce company in San Francisco reports, "We had a real problem because we were real newbies and however smart you are, you are still newbies. So it was a difficult team to back. They would ask, 'It's a great concept but how are you guys going to make it happen?' I think that is a lot of what investors look at." This situation is one in which venture capitalists draw upon their extensive know-who to access reliable opinions on the skills and work history of would-be founders of startups. As one venture capitalist in San Francisco puts it, "I do a tremendous amount of reference checking, I really spend a lot of time talking to people about people."

Reference checking, however, is tricky because the individuals who are likely to provide the most candid evaluation of entrepreneurs may not be the references provided to the venture capitalist. Investors value what people within their own network of contacts think about entrepreneurs more than someone they do not know. Thus, reference checking and other due diligence can be greatly aided by a strong network of local know-who that can be used to cross-check an entrepreneur's references. A Menlo Park venture capitalist describes the process.

> I want them to take me through their whole career from inception and then I'm constantly writing down names of people they overlapped with or might have overlapped with or whatever, and then I'm calling the people they didn't tell me about. What I'm looking for is someone I know and can call who will give me confidential information about this person's true performance and in exchange the tacit agreement or the implicit agreement is that I'm going to do the same for them when they call me about somebody. I don't believe in delegating reference checking. I think you've got to do it yourself because you need people you can really trust, this again goes to one of the advantages I have of being a long term Valley person and having a good network. I'm able to do incredible reference checks and know what it takes to make companies successful.

This emphasizes the importance venture capitalists place on their local know-who for evaluation and results in a system in which investment often follows connections and tacit knowledge flows. In the end, regardless of the track record of the management team, venture capitalists must invest based on their knowledge at hand. Because much of this knowledge is based on opinions, projections, and conjecture, which are extremely hard to codify, decisions often come down to a single individual's or several individuals' professional judgment. Von Burg and Kenney (2000, p. 1152) argue that a venture capitalist's decision to invest is based on "gut feelings about the people involved. Since these investments demand an envisioning process, there is a significant component of tacit knowledge in the investment decision that cannot be easily made explicit."

Although venture capitalists regularly indulge in revisionist history in which they claim to have been certain from the start that a particular company would be successful, the fact that only two or three of every ten investments are home runs belies such clear-cut decisions. Instead, venture capitalists act in situations of great uncertainty about the prospects of firms and where decisions have to be made quickly. To manage this risk, venture capitalists seek to gain as much knowledge as possible in a short time, resulting in a strong local focus in venture investing. This local focus becomes even more important once an investment is made and venture capitalists seek to support their portfolio companies and safeguard their interests.

Assisting Companies after Investment

The importance of proximity in venture capital investment is perhaps most marked after an investment is made, due to the active involvement of venture investors in their portfolio companies. Gompers and Lerner (1999, pp. 180–3) demonstrate that venture capitalists serving on a company's board are likely to be geographically proximate to the firm. They attribute this finding to the high transaction costs associated with visiting distant firms. Entrepreneurs who attempt to get nonlocal venture capitalists to invest in their companies often echo this sentiment. "My partner lived in Phoenix, Arizona and we were trying to locate the company in Phoenix and the VCs ⟨from Silicon Valley⟩ really hated that idea because they wanted to see us regularly."

This section analyzes three principal ways in which venture capitalists assist companies and outlines how each of these functions is facilitated by geographic proximity. While some activities do not necessarily require a physical presence, many venture capitalists and entrepreneurs value proximity for higher and more frequent interaction as well as assistance in which know-how and know-who can be transferred. Weekly and even daily phone calls and meetings are not unusual as investors protect their investment by providing the support of their experience and contacts.

General advice and strategy building

The most general way in which venture capitalists support entrepreneurs is providing advice on how to grow the company. Because venture capitalists have previously dealt with companies facing similar issues in growth, e.g., developing marketing campaigns, starting production, increasing sales, they are able to directly provide input or connect the company with someone who could provide this assistance. This process varies depending upon the number of companies with which a venture capitalist is working but generally involves larger issues of firm strategy rather than day-to-day decisions. A venture capitalist out of Menlo Park describes her relationship with entrepreneurs in the following manner.

> For some people it's just giving them a shoulder to cry on and a confidential ear to pour their hearts out once and awhile. It can be as simple as walking someone around the block who knows he has to fire a co-founder because, my god, it's hard to do. Sometimes it's giving them feedback after a meeting on how they handled things or how they could better handle a board member. I help them a lot with organizational philosophy – how to set clear lines of authority, how to compensate people and how to measure their performance – giving them tools for management. I find a lot of times these people are CEOs because they came up with a great idea not because they are good at management.

This assessment is mirrored by what many entrepreneurs hope to get from their investors. Although they may have significant business experience, they value the input available from venture capital investment versus another source of capital. One Internet entrepreneur in Redwood City argues that "It's great to have dumb money because, hey, money is useful but . . . it's not great to have people without a lot of experience to help you with the company." The level of interaction between a venture capitalist and an entrepreneur varies according to the stage of the company, the level of commitment of the venture capitalist, and the current challenges facing a firm, although almost all venture capitalists who invest in early-stage companies emphasize the importance of being close to their portfolio companies. One angel investor remarks, "I want to be able to see, feel, and touch the companies that I'm investing in on a regular basis. I obligated myself at least once a week and in practice I'm touching the company at least three times a week." Another investor based in San Francisco echoed this sentiment with his argument that "the way you add value is to be close. Face to face contact to trade gossip on what we know about the competition, popping in to brainstorm, going to board meetings, etc."

Thus, geographic proximity can and does play an important role in the level of support that firms are able to gain from venture capital investors. A telling example of this from the Internet industry is the way in which eBay selected its financing. Although eBay was actually generating a profit from the start, its founders wanted to grow the company quickly and were interested in outside financing. Despite receiving an investment offer from Knight Ridder that valued the company at $50 million, they decided to take a much lower valuation, $20 million, from Benchmark Capital since it gave them access to the venture capitalists' extensive contacts and experience (Stross, 2000, p. 29).

Setting metrics and accountability

In addition to advice and mentoring, venture capitalists also serve companies by setting specific goals and metrics for companies to meet and by holding managers accountable. This is a controversial aspect of venture investing because there is room for conflict between the entrepreneur and venture capitalists on the feasibility and relevance of various goals. Nevertheless, this goal setting and accountability provides firms with a clear timetable to meet and incentives to do so. This pressure from investors can in the best situations compel firms to grow with great speed. A senior vice-president with a business-to-business Internet company in Palo Alto compares the experience at his current venture-backed company with another startup at which he had worked that had no outside investors. He argues that the growth of his former company

> was limited to the founder's blinders. If he didn't believe in a technology or tactic, it stopped at him because there was really nobody for him to bounce ideas off of. There was nobody to force monthly metrics like,

Bam! Bam! Bam! Where are you? Where are you? Where are you? We grew at his conservative pace. There was very little accountability.

In contrast, he sees the venture investors in his current company as "driving results, driving definite accountability. They are in there looking at models, really really using thumbscrews. Not that the company is potentially failing, actually the contrary, but they want to be really in there." While some entrepreneurs may see this active role of venture capitalists as an unwelcome intrusion in their firm, many see it as a fact of life if a firm seeks venture capital. Because it is easier to monitor and meet with companies that are nearby, the intensity of this involvement corresponds with the distance between investors and companies.

In some cases the oversight of venture capital culminates in the replacement of the founder of the company and the hiring of a new CEO or president. While this potential loss of control may lead many entrepreneurs to forgo venture investment, other founders argue that this possibility is just par for the course. As the cofounder of an Internet software company observes:

> It's inevitable in going after capital. If you don't want to give up a piece of your company then you shouldn't go after capital. I think it's rare for founders to not lose control. I'm not really concerned about this. I've always been a believer that the intent is to grow and build a company that is really good. If I can bring people in to help towards that end, I think that's the best thing. I'm a shareholder just like the other investors and what you want is to grow your company and make it bigger and better.

However, not all founders are as sanguine as this one. Others have serious concerns with losing control of their company and are skittish about accepting too large a role for their venture capitalists. As the founder of one New York Internet software company forcibly argues:

> With VCs they'll try to throw out all your management and put in their own people. With VC management you get a feeling that these are the managers that the VCs rotate from one firm to another so they have more loyalty to the VC that keeps giving them jobs than to the company. I guess we have fair amount of paranoia about venture capitalists.

Partly as way of addressing this concern, this particular company selected venture capitalists from Chicago in an effort to limit their role.

Nevertheless, there is considerable acknowledgment on both sides that the skills that are important for founding a company are not necessarily the same as those needed to grow and manage a company. As the founder of an Internet services company in San Francisco notes, "What I like to do is notice interesting things that need interesting solutions and work on them. I am less interested in managing a company and it helps to have a CEO with strong contacts in the financial community and gray hair." Another founder of an Internet company located in San Francisco, who came from a largely technical background, recalls:

> We had three offers from venture firms and began serious negotiations with them and the first thing they all said was that they wanted to bring in as part of the deal an experienced CEO. So they introduced us and we liked her and went ahead and hired her as part of the deal. It just made sense. I didn't know or want to be CEO and as CTO I kept my equity and the company got bigger.

Thus, much like the role of proximity in increasing the frequency of mentoring, the use of metrics to hold the founders of companies accountable to their investors is supported by spatial proximity.

Making connections for companies

In addition to providing mentoring and metrics to portfolio companies, venture capitalists play an important role in introducing entrepreneurs and helping them establish relationships with sources for later financing, customers and suppliers, as well as a host of service providers such as executive recruiters and lawyers. This activity largely depends upon the strength of a venture capitalist's know-who and is generally strongest in the venture capitalist's local region. At the end of their book, Gompers and Lerner (1999, p. 327) acknowledge the importance of these local relations but do not directly address the issue of how proximity assists in the transfer of knowledge. Instead they focus largely on the reduction of transaction costs rather than on any recognition of the Marshallian processes of learning and variation that are also at work in these agglomerations. Furthermore, there is no analysis of the role of knowledge transfer within these regions. This contrasts quite markedly with the emphasis that venture capitalists and entrepreneurs place on the transfer of tacit knowledge and know-who to companies. As one Internet entrepreneur in the San Francisco Bay area put it, "We really want the guys who have the value-added in their rolodexes. We don't want just money."

This emphasis on the connections of venture capitalists ("the size of their Rolodex") was repeated again and again by entrepreneurs and venture capitalists. Venture capitalists emphasize that this "Rolodex effect" is not just limited to who they know but can be extended an additional layer to include the contacts of their contacts. As one venture investor explains:

> The truth of the matter is that it is not just us. It's all the people we know. We had a situation with one of our companies where the company started to stumble and we said what we need is somebody who knows how to do X. Well, no one knew how to do X but we had four people who had friends who knew how to do X and so all four of them went out and we found out who had availability of time and the interest and inserted them into the situation and "bing bing," things were fine.

The first and easiest set of connections that a venture capitalist brings are contacts and credibility within the venture investing community, providing avenues for future rounds of investment as the firm needs it. In the context of the Internet industry, these connections include valuable contacts with mezzanine investors and investment banks, which can provide the financing and services that help a company go public. Even before the period in which a firm considers making a public offering of stock, the connections of their investors helps firms get access to better-known venture capital firms that may not have been possible on their own.

Beyond connecting companies with sources of additional financing, venture capitalists play an important role in introducing their companies to potential customers, suppliers, and service providers. The ability to make key introductions and connections is what generally distinguishes "smart capital" from "dumb money" and for this reason is greatly valued by startup companies. This ability to connect companies with key partners is also central to how venture capitalists view themselves. One venture capitalist located in Palo Alto describes his and his firm's role in supporting companies as

> using our combined resources, skills and contacts to make the early stage companies happen which is where you really need those skills. Later down the road you can buy the skills but at the very early stage,

a deal with AOL means a lot to an early stage company and there's three or four people in the group who have ways into AOL or Microsoft or eBay or wherever.

Another important set of connections and assistance that venture capitalists provide for companies is recruiting senior management. While this can include the replacement of the founders and is often a worry of many entrepreneurs, an important task for young companies is finding skilled personnel to take on a particular task. One venture capitalist jests, "in the end we are glorified executive recruiters. I spend a lot of time looking at executive candidates and if they fit with one of the companies we're invested in, then we definitely introduce them." This search takes place largely through the networks that a venture capitalist develops in the course of his or her work and many times the source of leads on managers comes from deals that they evaluated but decided not to back. The founder of an Internet company from San Mateo was originally recruited to another startup by a venture capitalist to whom he had just pitched.

> The VC tells me, "You're an intelligent guy. You've got a great idea. It's not clear you have a business we want to back, we have some issues with your partner, but we have this investment that needs a VP of engineering, they're restarting the company, it looks like there is some equity opportunity, why don't you look at it."

Other entrepreneurs report less involvement by their venture capitalists in recruiting, although they still accord them an important role in the search process. The chief financial officer of a San Francisco business-to-business Internet company remarks:

> I was a bit surprised that they aren't more active on recruitment. We were hoping that they would pretty much walk in with a list of people that we could hire. That doesn't happen. They really don't have stables full of competent employees looking for jobs. What they do is add credibility to your business so that when I'm going out trying to hire a CIO, I can say that we're a Kleiner Perkins based company. That's of interest to someone versus a startup that has no VC money or second or third tier investment. They also help throughout the interviewing process because they'll interview the senior recruits. They've done it so many times across their companies that they understand what to look for and they look for certain traits that they think will add value to our company.

A final connection that venture capitalists help make for their companies are with a range of service providers such as accountants and lawyers. While lawyers and accountants are often a point of entry into venture capital networks, i.e., they can provide the introduction that first gets a venture capitalist to review a business plan, the opposite takes place as well. All these types of connections are based upon the know-who of venture capitalists and although this kind of contact is possible via phone, fax, and email, the general consensus among venture capitalists is that proximity provides clear and measurable benefits.

Brokering Knowledge

The ability of venture capitalists to assist successful Internet firms was dependent upon largely regional systems of personal contacts and networks (know-who) through which difficult-to-acquire knowledge about technology, companies, strategies, and markets (know-how) was created and quickly exchanged. Although in principle this process need not take place in spatial proximity, in practice proximity is often a central factor because of the largely tacit nature of the knowledge used.[4]

Venture capitalists' location in the center of a system of tacit knowledge exchange provides them with a great deal of hard-to-acquire know-how and know-who but they are not the only actors who possess these skills and connections. Others are well informed about technological breakthroughs and key players in industries possess the know-how to grow companies. Entrepreneurs themselves can make connections with suppliers, customers, and strategic partners, although it may take considerably more time to do so. What sets venture capitalists apart is their ability to speed up this process to a degree that provides their companies with a significant competitive advantage. As one Sand Hill Road venture capitalist argues, "What we're really selling is time. When you have a startup, time is your most precious commodity so you want to do anything that saves it."

The majority of the entrepreneurs also emphasized this need for speed. The founder of a business-to-business company in San Mateo argues:

> The one truth about e-commerce is that the last guy loses or perhaps the first guy wins. Like any big technology, it's going to create winners and

losers, it's going to create the haves and the have-nots. And at the pace of the Internet it's a land grab and you better be there first. We're after anything that gives an advantage out the gate.

Other entrepreneurs describe the situation as, "The name of the game today is GBF, Get Big Fast, ramp the company up really quickly, do the land grab, establish the beachhead. You now own the market and declare victory."

In the context of the Internet industry, in which companies were trying to quickly establish brands and market share, the advantage of speed that came from venture capital made it highly sought after and led to a clustering of the Internet industry around its principal sources. Nevertheless, it was not the amount of capital in a region per se that was the key element behind the rapid expansion of Internet companies, but the networks and knowledge associated with it. Regions with systems of early-stage venture capital in place had an advantage vis-à-vis other areas because venture capitalists were among the first to recognize the commercial potential of the Internet.

Foundation for
the Dot-com Boom

Regional venture capital systems do not emerge overnight but develop alongside and concurrent with the industrialization and development process. Crucial to their operation are the feedback loops that emerge as venture capitalists, entrepreneurs, and workers come together in various new ventures. Even if a new firm does not succeed, valuable information, experience, and contacts develop during the process. These new or strengthened connections within a regional financing system provide the basis for subsequent efforts to form innovative firms. If the new firms are successful, there are an additional number of valuable feedback mechanisms that emerge. The most basic result, the generation of new wealth, can give an added surge to the investing process.

The case of the San Francisco Bay region is emblematic of this, with a history of venture capital that has developed over the space of 40 years or more. As Kenney and Florida (2000, p. 123) note, "Venture capital in Silicon Valley was not created out of a whole cloth; rather it evolved gradually as an element of the endogenous growth of the region . . . As they became an institution, they also reorganized their environment." This history of venture capital in San Francisco and its influence on the larger regional milieu is central to understanding the formation and concentration of the Internet industry.

Regional Venture Capital Builds on
Earlier Industrialization

Venture capital investing as it existed at the end of the 20th century had a relatively short history. The first formal institutional arrangements for providing risk capital started in the 1940s and the limited partnership model that currently dominates the industry did not come into prominence until the late 1970s. Before the Second World War, entrepreneurs had limited access to capital except their own personal finances or customers and suppliers willing to extend credit (Wilson, 1985, p. 14). Firms that had significant startup costs associated with the development of new technologies and products often relied upon wealthy individuals for financing.[1] In contrast, it is estimated that as much as 80 percent of private equity investments are currently directed by venture capital limited partnerships (Fenn et al., 1995).

California has a long tradition of capital accumulation and re-investment, with roots back to the gold rush of 1849 continuing through a series of other resource extractions over the next century. As Walker (2001, p. 167) argues, "California is a compelling case of resource-led development. Its expansion to the present trillion-dollar economy was jump-started by a gold rush, maintained by a succession of silver and oil strikes, and sustained by long term extractions from farm, fishery and forest. Not until the middle of the twentieth century did the balance shift away from land-based activities." This historical backdrop, and most importantly the institutions and conventions surrounding production and reinvestment of capital, parallels the San Francisco Bay region's accumulation and investment of capital in the modern venture capital system.[2]

Although many accounts trace the origin of Silicon Valley to Frederick Terman's move to Stanford in the 1940s or the founding of Hewlett-Packard, the entrepreneurial activity within the broad scope of technology can be traced back to shortly after the turn of the century (Sturgeon, 2000). Industries and technologies such as wireless radio, vacuum tubes, television, and short wave were all present and active in the region from the 1910s to 1940s. Local wealthy businessmen, playing the role of angel investors, backed many of these earlier ventures.[3] Kenney and Florida (2000) recount a number of individual investors, such as Frank Chambers, Edward Heller, and Reid Davis, who were actively investing in small firms in the San Francisco Bay region during the 1940s and 1950s.[4]

The beginning of formal venture capital

Despite this level of activity, investment in early-stage companies in the USA remained an informal process for the first half of the century.

Even the example of the American Research and Development Corporation (ARD), which helped finance Digital Electronics, was not widely replicated. In San Francisco, the most organized investing activity was simply a loosely tied group of investors known as "The Group" or "The Boys Club" whose joint activities were largely limited to monthly meetings or one-on-one deals (Wilson, 1985). Partly because of this, but also due to a perception that there was a low supply of private equity financing for new companies, the federal government implemented the Small Business Investment Company (SBIC) program in 1958 that allowed investors to leverage their own money with federal dollars (Wilson, 1985; Bygrave and Timmons, 1992; Fenn, Liang, and Prowse, 1995). As elsewhere in the country, the SBIC program proved popular in the Bay area, beginning with the founding of the Continental Capital Corporation by Frank Chambers in 1959 (Kenney and Florida, 2000). Although the SBIC model dominated the scene during the 1960s, it dropped out of favor by the end of the decade and eventually was eclipsed by the limited partnership model.

One of the pivotal events leading to the ascension of the limited partnership was the emergence of Fairchild Semiconductor from Shockley Semiconductors. In 1957 Arthur Rock, an investment banker at Hayden Stone in New York, received a letter from Eugene Kleiner, one of eight engineers at Shockley Semiconductors interested in starting their own company. After meeting with the group, Rock and his boss agreed to find funds to invest in a new company.[5] Eventually Rock negotiated a deal with the Fairchild Camera and Instrument Company in which Fairchild loaned the group of engineers $1.5 million. In exchange it had the option to buy the company for $3 million in three years, with each of the founding engineers receiving 10 percent of the money and Hayden Stone receiving 20 percent (Wilson, 1985).

Although the terms of this agreement are quite different from what is now the industry standard, at the time it represented a considerable gain in engineers' ability to profit from their work. Additionally, because Fairchild Camera exercised its option to buy the company after just two years, it contributed to the spinning off of new enterprises as the original engineers, unhappy with the arrangements, again left to form new companies.[6] Encouraged by this success, Rock and Hayden Stone in 1960 arranged a private placement of $1.8 million in another California company, Teledyne, for 25 percent of the company, which also proved to be a profitable investment.

This experience, and the fact that a New York-based investment firm could compete on the west coast, convinced Rock that there was an ample market in California. Except for a few firms, there was not a great deal of risk capital in California at that time (Wilson, 1985).[7] Arthur Rock moved to the Bay region and formed one of the earliest and most influential venture capital limited partnerships with Thomas Davis. Davis was active in the northern California high-technology investment scene, but did not have many opportunities to invest. They began their own firm in 1961 based on $3.5 million raised from individuals, including many of the original founders of Fairchild and Teledyne. This trend of earlier waves of successful entrepreneurs and firms providing the capital for the next generation is a constant refrain in the San Francisco Bay region. Rock and Davis's first investment was $1 million in Scientific Data Systems (SDS), which was later sold in 1968 to Xerox for just a bit less than $1 billion (Wilson, 1985, p. 37). Later, in 1968 as many of the key employees at Fairchild Semiconductors left to start new companies, Robert Noyce and Gordon Moore approached Rock in search of funding to start the company that would become Intel and prove to be another success.

Rock and Davis's partnership was influential in the development of venture capital in the San Francisco Bay, for a number of reasons. First, their ability to earn 20 percent of the profits from investments was very attractive to other investment managers considering becoming venture capitalists.[8] Second, their reliance upon their own research and use of personal contacts for due diligence highlighted the importance of networks in evaluating and selecting deals. Third, they emphasized the importance of building companies rather than just lending money. Finally, the structure of a limited partnership allowed the mobilization of outside capital to be invested in local firms to which they could provide significant assistance. Although later this would be less of a problem, at the beginning of the 1960s San Francisco had something of a capital shortage for startup companies (Kenney and Florida, 2000).

While Rock and Davis's partnership was groundbreaking and highly influential, it was not the only venture capital activity taking place at the time. A number of other individuals and firms were also turning from SBIC models to limited partnerships.[9] The most famous of these early venture capital partnerships, Kleiner Perkins, was founded in 1972 by Eugene Kleiner and Thomas Perkins. Kleiner Perkins was also unique because Kleiner and Perkins were the first technology executives, as opposed to bankers or financiers, to become venture capitalists (Wilson, 1985; Kenney and Florida, 2000). This transition is the most visible sign of the iterative process through which venture capital investing creates the means for its own propagation and expansion through funding successful entrepreneurs.

Expansion and decline during the 1980s

The early 1970s were a time of expansion of venture capital offices in the Bay region, both through local growth and the introduction of branch offices of banks from New York and other financial centers. The scarcity of capital observed by Rock at the beginning of the 1960s had been reversed and in order to get access to deals early on and at favorable valuations, many east-coast banks chose to establish a presence in the region.

Although the mid-1970s was a slow period for venture investing, a reduction in capital gains tax and a clarification of rules on permissible investments by pension funds at the end of the decade brought about a surge of investment. Kenney and Florida (2000) report that in the four years starting in 1978 over 50 venture funds were raised by San Francisco venture capitalists, with a large percentage of the money coming from pension funds. Fueling this activity were the very attractive returns posted by venture firms like Kleiner Perkins during the 1970s and later the widely publicized success stories of venture-backed companies such as Genetech and Apple Computers.

As a result, the availability of venture capital greatly expanded and a number of relatively inexperienced venture capitalists entered the industry. Bygrave and Timmons (1992, p. 50) report that in 1983 half of the venture investors in the USA had three years or less of experience. This created a situation in which a great deal of capital was invested in competing companies in sectors such as hard disk drives which later in the decade saw a number of failures and shakeouts. The mid-1980s was also a time in which the due diligence process of getting to know managers of startups was compressed as competitive forces caused venture investors to speed up decision-making. These two trends are remarkably similar to what took place during the commercialization of the Internet.

The shift in the source of venture funds from private individuals to pension funds, which also took place at the beginning of the 1980s, resulted in greater importance being placed on performance benchmarks. This made venture firms with proven track records highly sought after and put them in possession of increasingly large funds. Thus, a venture capital industry that was relatively homogeneous at the beginning of the decade became much more differentiated at the end of the 1980s, with mega-funds, second-tier funds, and seed funds (Bygrave and Timmons, 1992, pp. 54–60; Kenney and Florida, 2000, p. 117). Because venture capitalists at the mega-funds needed to put larger chunks of money into play, they began to invest less at earlier stages, providing openings for angels and seed venture firms. The crash of the stock market in late 1987 ended the expansion of venture capital investing in the 1980s and contributed to a recession in the San Francisco economy. As venture capital reached its nadir in the early 1990s, investment shifted to different sectors such as networking and telecommunications, typified by companies such as Wellfleet (later acquired by Bay Networks) and Palm (later acquired by 3Com).

Renewal in the 1990s

In the early 1990s the infrastructure and networks put into place during the 1970s and 1980s helped top-tier venture capitalists to quickly act on their recognition of the commercial potential of the Internet. This history of venture capital played an important role since it was deeply embedded in a dense network of contacts that entrepreneurs could use to gain access. As one Internet entrepreneur notes, "If you live in the Bay area and don't have access to venture capital, you've got to wonder. Everyone knows a venture capitalist out here."

The region's industrialization also provided a rich system of angel investors who had become wealthy through earlier companies. Combined it created an environment in the early 1990s with many access points through which entrepreneurs could find capital. As one Redwood City Internet entrepreneur recounts:

> It's a great process. You know you're going to get the money and you know you're going to get it at the right terms, but until the first big angel signs up to the deal, you could be going months. You have no idea when you're going to close it and then you get the first big angel and then, it's all right. You're halfway there and it all comes together.

While securing venture capital is never a sure thing for a startup company, this entrepreneur's hubris is indicative of the venture capital resources available in the Bay region.

Another positive aspect of the region's history of entrepreneurial and venture-backed industrialization was the willingness, in many cases eagerness, with which people left secure jobs to enter the difficult and risky realm of entrepreneurship. While the ready supply of venture capital was a draw for would-be entrepreneurs, observing the high degree of entrepreneurial activity going on in the region was also an incentive. The founder of an Internet business-to-business company argues, "This place just breeds and encourages entrepreneurs. If you see your buddy Sam took a risk and took a company public and is a millionaire, you compare yourself and say 'I'm just as good as he is. He doesn't walk on water.' If you see that over and over again you say, well maybe I can do it." This environment also drew a number of entrepreneurs to the region, particularly early on when environments elsewhere were less attractive.[10] This combination of accessible risk capital and a supportive regional environment for risk-taking developed directly out of the area's history of venture-backed entrepreneurship.

Central to this process is a continuous recycling of entrepreneurs and capital back into the region and the creation of resources and knowledge geared toward funding, staffing, and expanding small startup companies. One entrepreneur characterizes it as, "kind of like the way people in Hollywood all have some connection to the movie industry. Here, everyone knows someone who's in the startup business and so has access to information on how to do it, people to partner with. The population just has a denser entrepreneurial body than probably any other place I've seen."

Thus, the regional conventions and supporting institutions contributing to the San Francisco Bay's dynamic entrepreneurial and venture-investing process go well beyond entrepreneurs and venture capitalists. Kenney and Florida (2000) and Cohen and Fields (1999) argue that many other actors, such as lawyers, accountants, and other service providers, as well as conventions like accepting equity in lieu of payment, emerged alongside the development of the region's venture capital system. This echoes the thoughts of a Palo Alto venture capitalist who argues:

> When you talk about the industry it goes beyond just what I've talked about. The accountants have to know how to deal with software companies, the lawyers, everybody. The whole industry here is focused around startups. You go down the street here and ask anyone what they do and they do something that is somehow related to a startup . . . They think of this as an investment opportunity. They make it happen.

This long-term and historically grounded base of venture capital investing illustrates Sturgeon's (2000, p. 47) argument that regional systems that support entrepreneurial behavior and innovation are put into place over the course of decades and defy quick fixes.

> Silicon Valley is nearly one hundred years old. It grew out of a histori-cally and geographically specific context that cannot be re-created. The lesson for planners and economic developers is to focus on long-term, not short-term, developmental trajectories. Silicon Valley was the fastest growing region in the United States during the late 1970s and early 1980s, but that growth came out of a place not a technology. Silicon Valley's development is intimately entwined with the long history of industrialization and innovation in the larger San Francisco Bay Area.

This review echoes Sturgeon's emphasis on place and the context created in a place's industrialization as opposed to a technological path dependency. The fact that the venture capital system within the San Francisco Bay region was embedded in a specific historical and institutional context presaged much of the dynamics of how and where the Internet industry emerged. Although neither the World Wide Web nor the first graphical web browser were developed in the region, the Mosaic development team was recruited to Silicon Valley by key actors in the region's venture capital and entrepreneurial system. This shift, both from academia to the private sector and from the Midwest to California, marks the beginning of the commercial Internet and the rise in importance of venture capital funding.

Beginning of the Commercial Internet, 1993–95

In 1993 when the future popularity of the Internet was still uncertain, the Mosaic development team, led by Marc Andreessen and Eric Bina, at NCSA at the University of Illinois made the Web a much more invit-ing place for the mainstream public. Tim Berners-Lee had created the necessary protocols for the Mosaic browser a few years earlier but had seen the Web primarily as a tool for researchers rather than a forum for the general public. By late 1992 Andreessen and Bina, along with a number of other programmers, created a software program named Mosaic that was released to the Internet public in February 1993.

Within a few months of its release, hundreds of thousands of copies of the program had been downloaded from University of Illinois servers and the Internet witnessed a surge of Web traffic.[11] With this rising popularity, including a front-page article in the *New York Times* business section in December 1993, the Mosaic development team and the NCSA experienced increased demand for user support and requests to license the software. The managers of the NCSA, who had largely ignored the original Mosaic project, began to take an interest in the future of the browser, slowly eroding the autonomy enjoyed in the initial phase of its development (Reid, 1997).

The rise of Netscape

By his graduation in December 1993, Marc Andreessen had grown disenchanted with the NCSA's approach to Mosaic, and left the Midwest for Silicon Valley where he took a job at a small software company in Palo Alto (Reid, 1997, p. 18). Although he toyed with the idea of starting a "Mosaic-type" business in Illinois, he did not follow through because he was unsure how to go about doing it and there was no one to ask (Reid, 1997, p. 21). In February 1994 he received an email from Jim Clark, the founder of Silicon Graphics (SGI), who was interested in starting a new software company related to interactive TV. Shortly before Clark left SGI in 1994, he was introduced to Mosaic and by the end of his first surfing session had emailed Andreessen suggesting that they meet. Although their first meeting focused on set-top boxes or a Nintendo-like product, they eventually decided to build a "Mosaic killer" and set up shop in Mountain View with $3 million of Clark's own money (Reid, 1997). The first order of business for the new venture, originally dubbed Mosaic Communications and later renamed Netscape, was flying to Illinois and recruiting the rest of the programmers who had coded the original Mosaic browser.

In retrospect, the potential of the World Wide Web and Clark's decision to back Netscape may seem obvious, but at the time the future of the Web was anything but a sure thing. Parallel to the development of Mosaic, Time Warner and a number of other companies were investing in the creation of a set-top box that would create a system of interactive TV (Lewis, 2000). Even those familiar with the Web were uncertain of its potential for business use given its reputation for anarchy and a decidedly anti-commercial atmosphere. One early Internet entrepreneur, who began his company in 1994, describes a high level of interest but low level of knowledge about the Internet on the part of venture capitalists in San Francisco.

Six months after graduating from Stanford [June 1993] we had our enhanced email and Web business plan together. Initially we walked

around Sand Hill Road and went to VCs who had been on campus . . .
we started calling ourselves professors, because what we were effec-
tively doing was educating the VC community on this new thing called
the Web. We would walk in the office, that was easy, everybody wanted
to talk to us . . . they wanted to know what we knew. Most of these
people did not even have a Web connection. A lot of these people
weren't even using email back then.

This unfamiliarity and uncertainty would soon give way to growing
interest as Netscape released products and expanded. Netscape's first
version of its Navigator browser was released via the Web on October
13, 1994 and was an immediate success, growing from zero to a 75
percent share of the browser market in just four months (Naughton,
2000, p. 251). Because it was significantly faster than Mosaic and
equipped with commerce-enabling technology such as secure socket
layer software, it quickly became the browser of choice for Internet
users. This, combined with favorable press coverage such as being one
of *Fortune* magazine's 25 Cool Companies in July 1994, dramatically
increased Netscape's visibility.

Nevertheless, as the founder of an Internet service company in San
Francisco remembers, it remained relatively little known outside the
circle of Web aficionados.

> In 1994 when I started at Haas (University of California-Berkeley's busi-
> ness school) the Internet was nothing, . . . Mosaic had just come out . . .
> but there was so little interest in general. Jim Clark came to speak at
> Haas in '94 and had an audience of only a 100 people . . . Most people
> didn't know what a browser was then – but good luck getting into a
> talk by Clark today!

Although Clark could have likely financed the entire company himself, Netscape turned to venture capitalists for funding in order to gain
access to their networks and receive help in recruiting senior management. Kleiner Perkins, led by John Doerr, bought 15 percent of
Netscape in 1994 and quickly recruited seven vice-presidents and a
CEO in just 150 days (Reid, 1997).

Early movers in the San Francisco region

Netscape was not the only Internet-oriented company in a region that
included the early activities of the founders of CNET, Architext (later
renamed Excite), I/Pro, Onsale, Hotwired, E-loan, and eBay. Perhaps

best known is the Web index and portal company Yahoo! started by two Stanford Ph.D. students, Jerry Yang and Dave Filo, in 1993. Although their web site was initially an effort to keep track of things that they enjoyed, their cataloging efforts addressed the fundamental problem of finding things on the Internet. The initial list, dubbed Jerry's Guide to the World Wide Web, was housed on a computer in their trailer office in a parking lot at Stanford University. At first, people found out about the list through word of mouth or email, but by the summer of 1994 the site was receiving tens of thousands of hits daily and in the fall of 1994 they had their first million-day hit (Reid, 1997).[12]

By the end of 1994, with the help of a friend in business school, the pair began work on a business plan. During the first months of 1995, they met with dozens of venture capitalists as well as receiving overtures from a number of companies such as Reuters, AOL, MCI, and Microsoft (Lardner, 1998). In April 1995 the pair accepted $4 million of funding from Mike Moritz of Sequoia Capital and began expanding the company. One of the first priorities was finding a CEO and, after a few months of searching, Moritz recruited Tim Koogle, a Stanford graduate with a wide range of management experience (Nocera, 1999). Recognizing their own limited experience with running a company, Yang and Filo were happy to pass on much of the day-to-day responsibilities to him while retaining the titles of Chief Yahoos (Reid, 1997).

A number of other Internet entrepreneurs and startup companies also emerged early on in San Francisco. Its history as an early node of the ARPANET meant that there was a great deal of familiarity with the Internet which, combined with the entrepreneurial environment of the region, generated a great deal of startup activity. The relative ease with which Netscape and Yahoo! were able to secure financial backing belies the challenge faced by many of these early entrepreneurs. Although the region had the venture capital infrastructure in place, the World Wide Web was still a big unknown and convincing investors of its commercial potential was the principal challenge. As the cofounder of an Internet services company remembers:

> For us the most frustrating thing was trying to explain about the opportunity. We were in a position where we had to describe what the Internet was before we could explain our business plan. We must have talked to 20 VCs who were bullish on the Internet in a weird abstract sense but when we wanted to show them our web site we discovered they weren't connected. It was crazy frustrating.

This issue of having to explain the Internet eventually subsided as more investors had the opportunity to surf the Web themselves. As an early Internet entrepreneur describes, "It took us about a year and a half before people began to understand what we were talking about. The first year and a half was mostly just talking, but by early '95, at least, a number of people had the opportunity to surf the Web and had some idea of what it was." Nevertheless, because it was unclear exactly what kind of business model made sense for the Internet, e.g., content provision and subscriptions were thought by many to be the only viable model, entrepreneurs were by no means assured of getting financing.

This learning curve and uncertainty was also present in the informal investing community, i.e., angels. Although San Francisco had a good supply of people willing to invest in small-scale companies, they too were relatively unfamiliar with the Internet. Later, as venture investing expanded, angels became increasingly organized to put larger amounts of capital into play. Nevertheless, as this entrepreneur reports, the system of organized angels was not developed in 1994.

> At the time [1994] the angel investor thing wasn't as mature. I would have to have been going around meeting people personally and roping money in 50K amounts. I really wanted to raise a reasonable amount of money so I wouldn't have to focus on money raising and could go build a product. But I couldn't go out and get a million bucks because no one had that level of confidence in the Web. I would have to go out and raise money in $50,000 increments and then I would have all these personal investors calling me on the phone every week that would just be maddening. So I got kind of scared off by that whole investment process and gave up on the idea.

The turning point for the Internet industry was the attention that Netscape's IPO drew.[13] Prior to Netscape the rule of thumb among venture capitalists was that a startup company needed to have at least four profitable quarters before going public. Clark, however, had pushed hard for Netscape to become a publicly traded company more quickly and on August 9, 1995 Netscape offered five million shares of the company at $28 a share. Demand was so great that the stock closed at more than double this price by the end of opening day. Netscape's performance in the public markets legitimized the Internet as a viable commercial space and imparted the idea that the Web was something that could make you rich in a few years. IPOs by Excite and Yahoo! in April 1996 further validated the Internet and increased the amount of risk capital available.[14]

First generation of Internet companies, 1996–97

After Netscape's groundbreaking IPO in August 1995 the Internet entered the mainstream. Microsoft, which had been developing its own proprietary network system, further legitimized the Internet with a day-long Internet strategy event in December 1995, that was followed by the announcement of the formation of MS-NBC. Although this increased attention was taken as validation of the early efforts of Internet pioneers, the size of the expansion over the next three to four years was difficult to imagine even for the most hopeful entrepreneur. As the founder of a San Francisco Internet company remembers, "We all had a sense that the Web was going to be something big but looking back nobody knew that it was going to be what it is today with billions of dollars of valuations."

However, uncertainty surrounding the rollout of the World Wide Web and how it would affect business tempered the growing recognition of its potential. Netscape had proven that Internet software companies could make money (at least through stock offerings), but no one knew if it would take hold in other industries. Places with a critical mass of small companies or people familiar with the Internet proved to be more fertile ground than large corporations for experimentation with commercial Web ventures. The founder of an Internet services company in San Francisco notes:

> Around January '96 I was still at Haas and went to a business conference on direct marketing. Over three days there were nine speakers, only two or three of these speakers even talked about the Net. This is January '96! And the context in which they mentioned the Web was, "We're going to sit on the sidelines and wait for infrastructure to develop." I went back to my hotel room and thought, a 50 odd billion dollar industry, at the time probably about 35 million Americans on the Net . . . there's got to be some way to bring the two together.

Armed with this sense of an opportunity overlooked by larger and more established companies, this entrepreneur returned to San Francisco and began to recruit his friends to start an Internet company.

The number of Internet-related IPOs tripled in 1996 and venture capital investment increased by 46 percent over 1995, surpassing the amount of venture capital money invested at its earlier peak in 1987 (constant dollars). Although it is difficult to get accurate numbers, Cortese and Hof (1995) cite a figure of $42 million invested in Internet companies in 1994; by 1996 PricewaterhouseCoopers figures indicate that more than $1 billion were invested in Internet companies.[15] Companies that emerged in 1996 and 1997 in San Francisco include Hotmail (which pioneered free email services), Healtheon (Jim Clark's second Internet company), Pointcast (a leading push technology company), and Webvan (an online grocer).

eBay was also founded at this time by Pierre Omidyar. Originally provided as a free service, the site became so popular that Omidyar was forced to move to a commercial server to handle the increased traffic (Stross, 2000). In order to cover the costs, he asked people to pay a voluntary service fee for items that sold. Although there was no enforcement mechanism, enough people who used the service sent checks that the web site was paying for itself from the start. Despite this early profitability, Omidyar wanted to grow the company quickly and for that reason looked for outside financing with strong contacts within the financial and business communities. This access to the know-who of venture capitalists was quickly put to work in the recruitment of Meg Whitman as eBay's new CEO. The company went public on September 24, 1998, and at the time had the fifth largest first-day gain ever in the history of the market (Stross, 2000).

The initially hot IPO market at the beginning of 1996 had slowed by the summer, creating some speculation that the Internet expansion was over, but after a slight dip in venture investing during the third quarter of 1996, the pace of investing quickly built back up. In fact, with the exception of the first quarter of 1997 the amount of venture capital invested in the USA increased every quarter until the market downturn in the second quarter of 2000. Although 1997 was a slow year for public offerings of companies, both Internet and otherwise, there were still rich opportunities for other exit strategies for venture-backed firms. Hotmail was a particularly influential example of the acquisition route when at the end of 1997 they announced the sale of a less than two-year-old company to Microsoft for close to $400 million.

Despite the increasing availability of venture capital during 1996 and 1997, it was still a relatively time-consuming task for entrepreneurs to secure funding. As one founder of a San Francisco-based company recounts:

> At the time you did whatever you could to get yourself some operating capital . . . Most of my work during this time was fundraising. Unfortunately, it still predated the "I have a business plan that's worth $100

million, why don't you invest $25 million in startup capital" of today's world. We were just looking to raise two to three million that would get us to a point, that we thought we'd be ready to raise a venture round.

This initial period of commercialization illustrates the advantage that accrues to firms and regions with the ability to move and adapt quickly to new innovations. The San Francisco experience demonstrates that this ability is built through a process of industrialization and incremental steps that lay the foundation for each subsequent round. As a result, firms within the region were able to move quickly when the opportunity of the commercial Internet emerged in the mid-1990s. With hindsight, however, it is clear that what was at first an advantage could and did quickly change into the driver of an unsustainable bubble of dubious companies funded by a hypercharged venture capital system by the end of the decade.

8

Panning for Digital Gold

The growth of the Internet industry during the last two years of the 20th century was extraordinary as measured by the number of firms and individuals experimenting with the Internet and the amount of risk capital investing in these companies. Nowhere was this more evident than in the San Francisco Bay region.[1] The situation contrasted sharply to 1994 and 1995 when only a few investors had begun to explore the possibilities of the industry. Venture rounds of $3–4 million were soon eclipsed by rounds of tens and even hundreds of millions of dollars as companies with no revenues or track records received astronomical valuations.

In retrospect, the situation was obviously absurd but reflected the exaggerated expectations surrounding the Internet and its ability to restructure the entire economy. The experience of the earliest Internet ventures and the seemingly unending desire for the stock of dot-com companies by the public markets further reinforced the process.[2] In large part, however, the firms formed during this time were built on little more than hype and hope and by 1999 the San Francisco Bay region was confronted with the ironic problem of too much capital and too little knowledge. Almost everyone involved "knew" that their particular investment or company would reach the winner's circle and spent accordingly. The result is that investment and new formation continued unabated until March 2000, when the markets could no longer sustain the process.

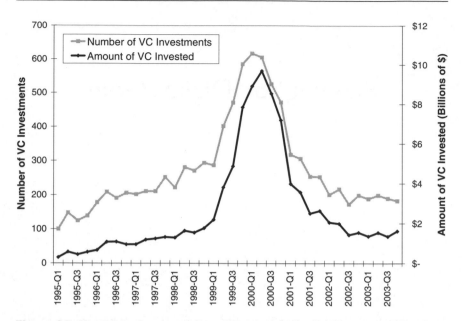

Figure 8.1 Number and amount of venture capital (VC) investments in the San Francisco Bay region, 1995–2003.
Source: PwC/VE/NVCA Moneytree survey.

A "New" Economy Fueled by "Dumb" Money

Figure 8.1 charts the growth of venture capital dollars invested in the San Francisco Bay region during the second half of the 1990s. Building upon its long history as the center of venture capital activity in the USA, the region was the location of approximately one-third of all formal US venture capital investment during the dot-com boom period. At the peak this translated into more than $9 billion of investment *per quarter*, representing more than twice the amount of venture capital that had been invested *per year* during the personal computing boom of the 1980s. The majority of this investment was in Internet-related firms and in certain fiscal quarters represented close to 90 percent of all venture capital investment in the region.[3]

The resulting atmosphere from all this investment in San Francisco contrasts sharply with the situation just a few years earlier. As one Internet company founder remarked:

The difference [between 1994 and 1999] is day and night. Now any dumb idea is getting funded. It has also helped us because it has made it a lot easier to get money. It is actually a little scary. What scares us is the frenzy to grab whatever and we just hope that there is a soft landing rather than a bubble bursting.

His assessment on the availability of capital is echoed by another early entrepreneur. "It took me a year to get funding [in 1994], but today it could take a month. It's just easier today because there is more money out chasing deals. The problem is that a lot of the money is dumb."

An embarrassment of riches

Despite the difficulty that the first Internet company had in maintaining its market share and turning a profit, the momentum started by Netscape's IPO continued unabated.[4] Venture capital investing really picked up speed in 1998 with a 33 percent increase over the amount invested during the previous year. This expansion was soon overshadowed by the 201 percent increase from 1998 to 1999 and the 83 percent increase from 1999 to 2000. Moreover, because of the successful IPOs of DoubleClick, Verisign, CDNow, and eBay in 1998, many Internet companies were positioning themselves to go public. After two years of relatively little IPO activity in 1997 and 1998, 1999 and 2000 saw a dramatic increase in the amount of venture-backed IPOs (see figure 8.2).

This capital availability put entrepreneurs in a unique situation in which there was more capital than viable businesses or experienced managers. This resulted in a compression of the due diligence process and high valuations for companies with dubious business models as venture capitalists competed with one another to get in on the best deals. While it never reached the point that "anyone" could get money, there were many opportunities for entrepreneurs who had the right connections and experience to start companies quickly. The firm Epinions is a good example of the speed at which companies were receiving funding. In the late spring of 1999 the company founders went from an initial concept to $8 million in seed financing in just 12 weeks on the basis of 16 PowerPoint slides but no budget and no product.[5] What the company did possess, however, was a core group of founders who had garnered a great deal of experience with some of the biggest Internet companies in the San Francisco Bay area,

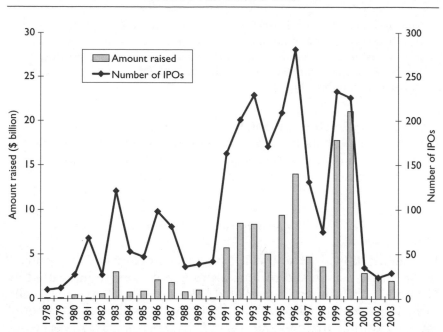

Figure 8.2 Number of venture-backed initial public offerings (IPOs) and amount raised, 1978–2003.
Source: Venture Economics.

including Yahoo!, @Home, and Netscape. This combination of know-how and know-who was very attractive to venture capitalists who were in frenzies to deploy money in companies as quickly as possible. Indeed, many were reported to have been irate at not getting into the deal (Bronson, 1999).

However, the supply of experienced entrepreneurs was limited even in San Francisco. Ironically, exactly as people with experience were becoming more difficult to find, the influx of venture capital was reducing the amount of know-how and know-who that startups received from their investors. Because venture capitalists were under pressure to put more and more money into play, they moved farther away from the advice and interaction that had previously character-ized their work (Zider, 1998). To counter this, venture capitalists in the San Francisco Bay area out-sourced many tasks, such as executive recruiting, to the region's supplier networks and increased their use of virtual CEOs or entrepreneurs in residence who could be pressed into service on behalf of portfolio companies. As one venture capitalist in San Francisco remarks, "It has really changed in the past few years. Because venture capitalists have so much money, they really don't have the time to spend with companies that they did before . . . my gut tells me that they probably have more companies that they can work with."

Constantly changing strategies

Another constant refrain during the commercialization of the Internet was the continuous evolution of what was considered a viable business model. A telling example is the San Francisco company PointCast, a leader in push technologies. In 1996, the use of push technologies to deliver content, news, and other information to computers was considered cutting edge, but three years later in the midst of a boom of venture investing PointCast was running out of capital and was eventually sold for $7 million (Himelstein and Siklos, 1999). This rapidly shifting focus could also prove beneficial, provided a company fit a model that was coming into favor.[6] As the founder of a business-to-business company remembers, "We started putting the business plan together in January '98, had our first meeting with a VC in April. We got in at the right time. A year ago no VCs were interested in services for old-line industries. Later on in the year, peoples' understanding of business-to-business e-commerce changed."

The shifting of business strategies is also evidenced by the creation and consolidation of web portals in 1998. America Online was the leading competitor but other major media companies, such as Disney (which acquired Infoseek), Yahoo!, and Microsoft, all pursued the same goal of being the biggest destination site that offered users an array of products and services. Many companies leveraged high market capitalizations to acquire smaller companies (such as the acquisition of Hotmail by Microsoft and Snap! by NBC), bringing them under their umbrella to offer more features to attract surfers. Other deals were put in place, including Excite's agreement to pay Netscape $70 million in exchange for a prominent placement within the Netscape site (Green, Himelstein, and Judge, 1998).

Based on the success of consumer sites such as Amazon.com, eBay, and E*Trade, venture capitalists in 1998 focused on retail sites on the Web that sold such things as furniture, prescription drugs, pet supplies, and groceries. This marked a shift from companies with a technological edge to an emphasis on marketing and sales, as shown by Quokka Sports, Wine.com, and Beyond.com. Some entrepreneurs even distanced themselves from the traditional technology rubric enjoyed by San Francisco companies. As the founder of an e-tailing company argues, "We want to be as little techie as possible. We're not a technology company, we're a retail company. Technology for us is an enabler of what we try to do, which is retail goods."

Because the barriers to entry to these marketplaces were relatively low, companies focused on establishing substantial market share fast enough to discourage competitors and venture capitalists committed to spending large amounts on marketing and sales. A number of San Francisco venture capitalists competed with one another in a wide range of retail products. While a similar rivalry took place at the start of the Internet industry, e.g., Kleiner Perkins backing Excite and Sequoia backing Yahoo!, many rival e-commerce companies were created expressly as a reaction to the announced funding by other venture capitalists. Moreover, these startups were funded at much higher levels than earlier Internet companies. For example, in the online drugstore market, the two big rivals, Drugstore.com and PlanetRx, backed by Kleiner Perkins and Sequoia, received hundreds of millions of dollars versus the relatively small amounts invested in 1995.[7]

Factors Behind the Boom

Although the dot-com boom of the late 1990s is likely to go down in history as one of the great bubbles (comparable to the tulip craze in the Netherlands in the 16th century), there were several compelling factors behind the willingness of normally rational people to behave with such "irrational exuberance."

Internet would change everything

One of the most important contributing factors to the dot-com boom was the hype surrounding the transformative nature of the Internet. While arguably less important to the economy than the introduction of the electric engine at the beginning of the century or even air conditioning, the Internet was touted as a way to completely revamp business models.[8] Much less attention was paid to the amount of time that it took for these innovations to change the structure of businesses and industries (David, 1990). As misguided as these ideas are in retrospect, they had a powerful influence on the entrepreneurs who were starting these companies. As the founder of a search engine site observes, "We all tended to believe the BS. We thought we were in the new economy when we were actually in a bubble. I still believe that the Internet changes things fundamentally but we used to think that everything would instantly be on-line. Now that idea just looks silly." Another Internet company founder echoes this sentiment and notes that there were a number of clear benefits promised by the Internet.

> We had convinced ourselves – and the world had convinced itself – that this was a whole new paradigm. We were not alone in believing that the world had permanently changed. The Internet and this new age of global communications would provide insight on customer demand, manufacturing inventory and would eliminate the huge swings in the economy caused by over-inventory. The collective wisdom at the time was that we would have perfect insight into customer demand because of the Web.

These sentiments were widespread and given this belief in dot-coms' ability to transform markets, traditional measures of companies' viability were largely ignored. Rather, metrics on the speed at which a company acquired market share and its ability to attract attention from users and the press were emphasized. As a venture capitalist in Menlo Park explains:

> What typified the last 18 months of the boom was that regular economic terms were no longer used to value companies. There was fierce competition to do deals which led to leaving behind traditional metrics. Valuations of companies were made under the assumption that they would continue to be priced above what traditional financial models would suggest. These valuations used market share or eyeballs to judge companies. It seems ridiculous today but those were the measures at the time.

Because standard valuation tools were no longer being used, entrepreneurs and dot-com companies had considerable incentive to chase market share regardless of the cost. As the market and investors returned to a concern for profitability in 2000, companies that had been designed with growth as a maxim were hard pressed to change. The cofounder of a retail-oriented company notes:

> Our investors kept telling us that we could always get more money as
> long as we spent reasonably and increased market share. The told us
> not to worry about profitability. So it was if we got outfitted to play
> hockey and although we played a good game of hockey, the game
> suddenly changed to football and we looked silly with skates on.

Clearly, one can look back at this time and critique companies for
spending so extravagantly but it was extremely difficult for a company
to both raise and husband risk capital during this time. Although there
were a few contrarian voices amidst the market madness, it is not clear
whether the "sane" voices could be heard over the din of a bull market
(Cassidy, 2002).

Slow growth was not an option

Historically, an IPO was limited to companies with four quarters of
profitability and used to raise money to execute carefully planned
strategies for expanding existing product lines into new markets.
During the dot-com boom, an IPO almost became a goal unto itself as
companies attempted to prove that they were the dominant player in
their market niche. With this drive it is not clear if a company could
have gone slowly. The chief marketing officer of an e-tailing company
notes:

> Almost before we could get the building blocks in place there was pres-
> sure from the board and CEO to get big fast. It was a land grab. Lots of
> opportunity, lots of money but it was over-hyped and over-charged.
> Lots of basic business ideas got thrown out the window. For example,
> we never built the value proposition that the brand would represent
> . . . because . . . there was no time to do the necessary research.

Without time, companies simply concentrated on grabbing as much
publicity as they could. Another Internet company founder remembers, "Part of the problem was that you had to take your company
public as a public relations and marketing device. Yahoo! got an enormous amount of recognition because of their IPO. Since everyone else
was going public it was hard not to and still be taken seriously."
    But perhaps most important to the survival of companies was
raising the financial resources necessary to defend market share from
similarly or better funded competitors. The founder of an informa-

tional portal company reports feeling compelled to pursue further rounds of fund-raising simply as a defensive strategy.

> By the end of 1998 we started thinking about how we could compete with the Disneys and the Amazons. We still had money from an earlier round but felt that we needed to bulk up to withstand the competition. At some level if everyone else has hundreds of millions and you have single digits you will lose out. We were on our own in a huge market with lots of dollars that would easily swamp us.

Others describe a similar defensive mindset behind raising funds: "You wanted to build up a war chest. You saw your competitors with hundreds of millions who might not have been as good but still could have out-spent you and driven you out of business."

The end result was a large influx of capital without much oversight or direction. Money was spent, market share was garnered, publicity was gathered, but despite these temporary successes many dot-com companies were unable to transition into lasting business models. As one senior manager notes:

> What the campaign lacked in strategic foundation it had in good execution. It was insane. A frenzy to establish yourself as the leader. Lots of dumb money and unrealistic expectations of growth. It would have been different if there hadn't been other companies in the same space but that was the thing. Without the competition we could have taken our time.

With multiple entrants in a wide variety of markets, however, there was little incentive to move cautiously and spend money slowly and multiple incentives to try to become the biggest company in a particular space. Unfortunately, this inevitably led to cut-throat competition in which companies with competing war-chests successfully drained the resources from one another until the survivors emerged with few remaining capital resources with which to create viable companies.

Avarice and ambition

While competition lies at the heart of a capitalist economy, the frenzy of equally well-funded competition resulted in a lot of money chasing competing, and in the end unsustainable, business models. In large part this was due to another hallmark of capitalist development, avarice. Avarice to strike it rich individually but also ambition to create a company that would dominate a particular market segment.

Although financial return has always been an important part of entrepreneurialism, it has not traditionally been the sole motivating factor in the San Francisco Bay region. Many engineers and company founders in earlier decades have been equally if not more motivated by the technological challenge that their companies and products confronted (Saxenian, 1994).

However, with increasingly short time horizons to IPOs, the start-ups which comprised the dot-com economy were viewed as investments that would produce astronomical gains in the short term. This observation from a Menlo Park venture capitalist mirrors those reported by a number of risk investors.

> We saw a panic among the investors to get as much money to the market as possible. The market was in a frenzy. At the height it was so easy [for a venture capitalist] to get liquidity, it was all about deploying money. People thought they could make easy money and greed really took over. There were some fundamental technological events but the liquidity in the market attracted a lot of people. But the fact that the bubble was going to come down was not a matter of if, it was a matter of when.

This quest for easy money was widespread throughout the early investment community, shareholders, managers, and employees of dot-com companies. At the height, the size of a company's potential upside became essential for attracting and retaining employees. As a company founder argues, "Because everyone was sloshing in money it would have been enormously difficult to maintain management and staff if you didn't say you were moving towards an IPO. People would have left for other jobs."

The effect of this greed was pronounced partly because there were very few countervailing attitudes. It was not, however, simply a matter of irrational exuberance because at the individual level, investment decisions were rationally made in response to market signals that indicated a demand for these companies at extremely high valuations. Many people amassed sizeable fortunes during this time following this investment and employment strategy. A former dot-com CEO contends, "The dot-com boom was part bubble, part irrational and part individuals doing rational things that when combined with everyone else was irrational. We ended up with a situation where all the independent decisions being made added up to the bubble."

In the end, it was the wide availability of capital at all stages of investment that provided both the incentive for the foundation of dot-com companies and their explosive growth at any cost.

Too Much of a Good Thing

Thus, in 1999 and 2000, the San Francisco Bay region was confronted with the unusual problem of "too much" capital. The responsibility for this over-deployment of capital is diffuse and in many ways is shared by everyone involved or invested in a dot-com company. This analysis, however, is not interested in identifying particular parties as the "culprits" but in assessing what took place and its impact on the region. One factor for this stance is the rough consensus observed within the entrepreneurial community that responsibility for the bubble goes well beyond any individual or set of actors. A typical argument made is that "At the height of the bubble it was driven by everyone. People thought it was a free ride. It's like a gold rush. It looked like you couldn't miss. That's why I joined. Five years ago it just looked like it would go on forever."

From the entrepreneurial perspective the availability of cheap capital was an opportunity to fuel a company toward fast growth. While there are examples of entrepreneurs who purposely misled investors, in large part this pursuit of capital was predicated on the belief that the Internet fundamentally changed business. By their very nature entrepreneurs must be unreasonably optimistic and explain why the numerous stumbling blocks will not derail their company. As one company founder explains, "Entrepreneurs are supposed to be the gun-slingers. We're supposed to go out and make it happen."

In this role, entrepreneurs were delighted with the access to venture capital rounds at high valuations which gave them considerable resources without the loss of equity. This perspective also highlights why capital was so easy to spend. As the founder and CEO of an early retail-oriented company argues:

A good entrepreneur looks for what's cheap and looks to see if they can replace what's expensive with what's cheap. In the go-go era, capital was cheap. People used capital to replace what was expensive . . . to make brand awareness, to hire people. It looks wasteful today but in the boom, you could get millions of dollars for just a small piece of your company.

Venture capitalists are an attractive group to blame given their direct role in providing hundreds of millions of dollars to companies, their high visibility in promoting dot-com firms, their considerable returns on certain early investments, and their quickness in discontinuing capital commitments to companies once the public market turned.[9] Even so, while venture capitalists were central to the dot-com boom and bust, a whole range of actors were also involved and self-interested in the promotion of Internet companies. A former company president suggests that "It is simplistic to point fingers of blame at any core group. If you want villains, you need heroes but who was saying no? Who was the hero within the Bay area saying that we should do otherwise? There wasn't anyone."

The biggest problem is that the flood of risk capital significantly altered the investment patterns of many venture capitalists. Traditionally, venture capitalists have served as "technological gatekeepers" who examined new technologies and businesses carefully and only invested in those with the best potential (Florida and Kenney, 1988c). During the boom, however, this gatekeeping role was diminished as more careful investment decisions descended into a rout of chasing companies to invest.[10] This is central to the dot-com bubble. The availability of capital provided incentives for people to start, relocate, and join dot-com companies in the San Francisco Bay region and used metrics based simply on how fast they grew. A long-time Silicon Valley entrepreneur remembers:

> The two types of people who should have known better are the VCs and the public equity analysts who overstepped the bounds. These are smart rational people who have been in the market for a long time and were pouring money into companies. You suddenly have these people who knew better, raising a $100 million fund, $400 million fund, $1 billion fund. An entrepreneur is like a wound-up puppy dog who can't objectively look at the company but that is what VCs should be doing. That system broke down.

In short, the dot-com era descended into the opposite of traditional venture capital investing, i.e., adding value in the form of know-how and know-who. Rather, investing in dot-coms was about putting as much money into play as possible with little of the value-added that traditionally accompanied it. The founder of an Internet infrastructure company observes, "I wish the VCs had been more VC-ish and acted less like some big bad funder dudes who were managing investments. They should have added more smarts to the money they were passing out because it really was just dumb money."

Thus, despite this history of more reasoned investments, many venture capitalists distributed vast sums of money to companies with very questionable business models. While some blame can be traced to the inexperience of new entrants without the perspective and caution of long-time venture capitalists, it did not exclude the well-established firms. This is because despite the questionable business models and high risks, real money was being made by investors in Internet companies. Venture capitalists are risk-takers by design, and in a standard portfolio of companies they expect that 20–30 percent of their investments will pay for the rest. Venture capitalists knew that the investments were risky but were inclined to go forward despite the fact, because market signals continued to point toward the potential for a good return. As a Menlo Park-based venture capitalist argues:

> Private equity is a high risk, high beta game . . . it's not for the weak of heart. Now it's easy to look backwards and go "What was I thinking?" It is easy to handicap the market in hindsight but real money was being made during this time. I think it is human nature to look for someone to blame. I'm sure that there is some criminal actions but I think it was largely part of the very human process of following the money. Lots of people got to participate in the upswing and now we're in the downswing.

While regrettable in hindsight, the cause of the bubble in investment was not a matter of deception, manipulation, or corporate misconduct. Rather, it was the outcome of a new technology that for a time seemed to be in the process of restructuring the entire economy, providing small companies with a golden opportunity for growth. In retrospect, the idea seems absurd and even during the late 1990s people openly debated the extent to which the dot-com economy was a bubble. Investors and entrepreneurs, however, by their nature find it easy to see the folly inherent in other companies and investments but believe that their individual experience will prove the exception. Whether it be as virtuous as a belief in the quality of the technology of the company or its business model or as mercenary as the greater fool theory, it drove continued investment until the public markets could no longer sustain the demand for new IPOs and dot-com company valuations.

Bursting the bubble

In the summer of 1999 the public markets experienced what many people thought was a long overdue correction for highly valued dot-com companies. Between July 9 and August 4 the Goldman Sachs Internet Index dropped by almost 30 percent and the NASDAQ dropped by 9 percent (Sparks and Laderman, 1999). Nevertheless by mid-August many of these stocks rallied and by the end of the year had regained or surpassed their pricing before the slump. Many Internet companies went public during this time including Webvan, whose IPO on November 5, 1999 took place just six months after it started commercial operations.

The NASDAQ index, where many of the public Internet companies were listed, closed at an all-time high of 5048 on March 10, 2000. The NASDAQ began to drop significantly in the middle of April and by May 2000 was over 38 percent off its March peak. The public markets rallied through the summer but in November 2000 experienced another drop and continued to decline through the first months of 2001. By April 2001, the NASDAQ was 68 percent lower than its peak 13 months earlier.

The upward trend of venture capital investing that had started in 1997 came to a close after the second quarter of 2000 and for the first time in three years began to decline. This downturn hit a number of Internet companies very hard, particularly those that had leveraged a great deal of capital or were focused on retail sales. Hundreds of Internet companies have gone bankrupt, tens of thousands of dot-com employees have lost their jobs, and the San Francisco Bay region saw the unemployment rate rise by over 4 percentage points in the space of little more than a year. While the dislocation and hardship experienced by individuals is very real, it is also clear that the system surrounding the bubble in dot-com companies was unsustainable and its eventual bursting was a question of when rather than if.

Despite many people knowing better, companies were begun with unhealthy cost structures that could not be maintained without an environment of millions of dollars in risk capital and public markets eager to buy shares of companies without profits, clear revenue streams, and good future prospects. As a result when the tide turned, many dot-com companies found themselves meeting standards that were no longer considered worthwhile metrics. As a Palo Alto venture capitalist explains:

If all these valuations were predicated on a market where profit was not a determination of value, you had to realize that it was going to change. Companies wanted to grow on the metrics that mattered so there were very unhealthy burn structures. When they had to meet new metrics it was an extremely fast turn and not all companies could make it.

The challenge then became reworking companies to be in line with the back-to-basics orientation of the changed market. This was a decidedly difficult task for many founders and managers of these companies who had grown accustomed to easy money or had extremely marginal business plans. As a result dot-coms instituted a series of layoffs and cut-backs in desperate attempts to stay in business but in many cases these were insufficient to hold off bankruptcy.

Bankruptcies, layoffs, and acquisitions

Obtaining accurate figures on Internet industry layoffs is a difficult task given the wide range of definitions for these companies and the lack of authoritative data on recent economic trends. Due to these issues a combination of private and governmental data sources are used to construct a composite image of the amount of job loss and bankruptcies in the USA and San Francisco Bay region. Because these private data sources do not systematically survey companies but track those which have achieved some level of visibility, these figures undercount the number of jobs lost and company closings. For example, a self-financed dot-com company with two employees that closed before producing a marketable product would not show up in these figures. Nevertheless, this does provide a reasonable overview of the size and scope of the bust.

Despite the non-stop media attention on dot-com companies during the late 1990s, the actual number of firms and employees was relatively small compared with the overall economy. At the height of the boom, estimates of employment ranged from 180,000 based on Hoover's Online data to 726,000 from the University of Texas Internet Indicators project.[11] Even the high end of these figures suggest a relatively small sector in employment terms which was vastly outweighed by the media attention surrounding it.

Thus, it is not surprising that the overall number of lost jobs directly attributable to dot-com companies is relatively low. One of the best data sources that tracked this at the national level is shown in

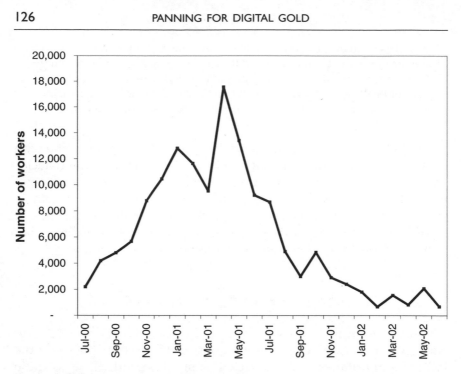

Figure 8.3 Dot-com worker layoffs in the USA.
Source: Challenger, Gray & Christmas.

figure 8.3. These data are assembled from figures obtained from news
articles and press releases on dot-com layoffs. This suggests that
nationwide the number of dot-com layoffs was in the order of 142,000
as of June 2002. More significant is the pattern of layoffs shown over
time. Initial rounds of layoffs began in the spring of 2000, approximately three months after the public markets began to decline, and
continued over the next 18 months with particularly strong peaks in
early and mid 2001. This represents the efforts of companies to reor-
ganize themselves through large layoffs to cut costs and lower their
burn rate.

Data from the Internet industry trade magazine *Industry Standard*
shows that 34,200 jobs were lost to California-based companies from
December 1999 to August 2001 (see figure 8.4), which represents
approximately 36 percent of dot-com job losses for the USA.[12] As in
figure 8.3, the bulk of layoffs cluster in the beginning of 2001 with the

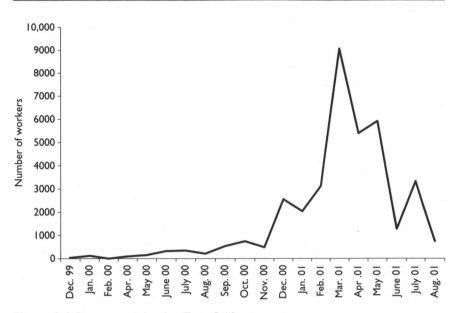

Figure 8.4 Dot-com worker layoffs in California.
Source: Industry Standard.

trend tapering off in the summer of 2001. Unfortunately, in August 2001, the *Industry Standard* joined the statistics of bankrupted companies, ending this very useful data series.

Figure 8.5 outlines three different sources detailing the total number of dot-com company shutdowns in the USA. Because each source relies upon its own definition, the exact numbers differ from one to the other. Nevertheless, all three sources show a similar temporal pattern as evidenced in the data on dot-com layoffs, with the bulk of company shutdowns occurring at the beginning and middle of 2001 and a steady trailing off in numbers. Again, determining the exact makeup of these companies is difficult but Webmergers reports 78 and 160 dot-com shutdowns in all of California for 2000 and 2001 respectively. Of these, more than 70 percent were located within the San Francisco Bay region.

Because the figures presented so far are relatively small and from private sources that do not attempt to include all companies, it is useful to compare them with governmental figures on unemployment.

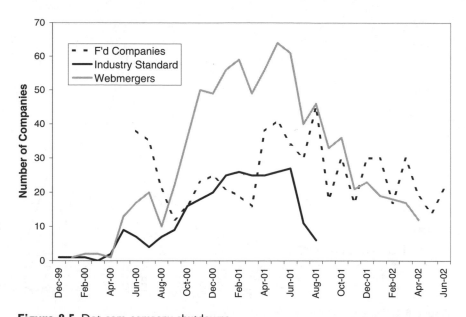

Figure 8.5 Dot-com company shutdowns.
Source: based on data collected at the following web sites: http://www.fuckedcompany.com (see Kaplan, 2002), http://www.theindustrystandard.com and http://www.webmergers.com.

While these data are not specifically focused on dot-coms, they likely reflect the multiplier effect of these firms on the entire economy. Moreover, because dot-com companies did experience large numbers of layoffs during this time, it is very likely that a major percentage of these job losses are associated with dot-com layoffs either directly or via multiplier effects as companies and individuals curtailed spending.

As figure 8.6 outlines, the unemployment rates within the San Francisco Bay region had been trending lower throughout 1999 and 2000. With the coming of 2001, however, unemployment quickly began to rise throughout the region and at a considerably faster rate than the state of California as a whole. Particularly hard hit were the counties of Santa Clara (the heart of Silicon Valley) and San Francisco, which had been central to the dot-com boom. In June 2001, Santa Clara county's unemployment rate became higher than the state average for the first time since 1983 when the online EDD (Employment Development Department) dataset began.

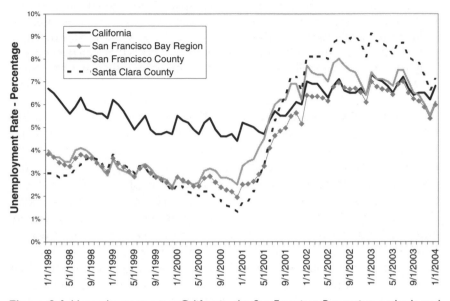

Figure 8.6 Unemployment rates: California, the San Francisco Bay region, and selected counties.
Source: Employment Development Department.

End of the venture capital model?

One of the effects of the dot-com boom was the development of extremely skewed expectations of what employees could anticipate from their jobs. Many people came to view a 12-month stint at a dot-com company as the means to easy wealth and an early retirement. While it is certainly an asset to entrepreneurial companies to tap into a labor market that is willing to work at startups, this type of expectation, i.e., high reward with little risk, was untenable. As a result, and coupled with the high demand from other dot-com startups, companies were formed around increasingly less qualified core people who nonetheless expected to reap big rewards.[13] One long-term technology executive observes:

> This had been a very unhealthy environment, particularly in the Valley. We had created expectations among employees that were unsustainable. The view was that everyone should be worth a million bucks after a

year. And if they couldn't get that in company A they would go across the street to another company. Hiring people was virtually impossible. We were paying these incredible bonuses – up to $5000 just to introduce someone to the company – it was bizarre. The irony is that a lot of people left us to join these IPOs and now they are gone while we are still around.

Silicon Valley has a long and rich history of rewarding risk-takers who start new companies and expand the technological horizons. Dating back to the founding of Fairchild Semiconductor in the 1950s, the Silicon Valley model pioneered the ability of engineers and entrepreneurs to capture a great deal of the rewards of their work. One of the essential parts of this working well is a risk and reward structure that is perceived as fair and equitable. In other words, those with talent and a willingness to work hard had a fair chance of reaping rewards. During the boom, however, the system of rewards became increasingly arbitrary, as business models were funded and successful IPOs seemed to be more about timing and public relations than the fundamentals of the business. This level of arbitrariness led to many perverse responses. As a dot-com founder explains, "At the pinnacle of things people were maddened by the crowds. You saw a lot of bozos that just happened to work at the right company at the right time be worth millions and that made people crazy. That ended up poisoning someone's outlook and made a lot of them cynical about what was important in a company."

Once it was felt that the rewards were based on luck, much of the entrepreneurial activity became an exercise of chasing dollars and raising the largest IPO rather than focusing on the basics of a company. Ironically, it was often the people who were most committed to the company who benefited the least from the rise in stock prices because they were committed to the long-term prospects of the company.[14] An Internet company CEO observes that contrary to the traditional arguments for stock ownership, often those who sold early reaped the most financial rewards.

> The worst thing was the injustice in who made lots of dough and those who didn't. Those who sold the stock made lots of money and those who held on are at a much lower valuation. In retrospect one of the best things you could have done for someone in 1999 was to fire them because they would usually go off and sell their stock in a fit of pique. In the end it was pretty random who benefited and some of the least deserving people made the most.

It is the randomness of the rewards that is most troubling for the long-term health of the San Francisco Bay entrepreneurial model. When it becomes less of a business investment and more of a short-term gamble, the markets will reward those who seek to game it and entrepreneurs who form the core of the new business innovation will see a system stacked against those who would proceed in a manner most beneficial to a company.

Moreover, the focus on the quick return shifted private equity capital investors away from companies that might have less spectacular short-term returns but stronger long-term prospects. This makes perfect sense at the individual investor level since they are maximizing their short-term returns; however, at a systemic level it makes it harder for more traditional Silicon Valley firms to raise funds.

> No one wanted to fund us because we weren't a dot-com. I met with 30 VCs and they wouldn't fund us. We got our initial money from people who didn't believe in the dot-com world. Our valuations were fractions of what equivalent dot-coms were getting and it was very frustrating. But after the dot-com crash people were looking for companies with customers, product and revenue. In January and February 2001, I started to have two to three meetings a day with VCs. It was like a switch went off and suddenly people wanted to talk with us. It was a great position to be in.

Although this entrepreneur's experience eventually turned once the dot-com bust had begun in earnest, his experience is emblematic of the larger issues taking place during the late 1990s. Simply, the traditional focus and model of technological development in the San Francisco Bay region had shifted to the dot-com model. A senior venture capitalist describes it thus:

> Silicon Valley went from selling technology to Wal-Mart to the business of using the technology to kill Wal-Mart. Companies started trying to sell to consumers themselves. Everyone can identify with these companies. The press writes about the stuff and the reader identifies with them because they can grok them. We went from zero to thousands of consumer-oriented companies. The boring startups, the companies that sell technology directly to companies, were still going like the little engine that could, and no one paid them much attention. We've spent a tremendous amount of money and effort to show that dot-com consumer businesses are pretty bad businesses . . . they were very different from what is at the core of Silicon Valley.

This shift was at the center of the dot-com boom and reflects that while the venture capital system of the San Francisco Bay region is very efficient at combing capital, knowledge, and labor to form new companies, there was a disconnect from the type of knowledge and business models that had traditionally been the source of its growth and the type of companies that formed at the end of the 1990s. In short, the knowledge and technology that was being leveraged through capital and labor during 1998 and 1999 was not particularly useful to have.

9

Dot-com Hangover?

While the dot-com era is commonly (and to a large extent rightly) perceived as a feeding frenzy of capital-chasing dubious business plans, it is reductionist to limit the analysis of this recent era thus. As tempting as it is to stereotype the Internet industry as 20-something CEOs wasting millions of dollars on Superbowl ads, expensive office chairs, fussball tables, and parties, it provides little insight or learning beyond the obvious truisms of basic business fundamentals and cautious investing. In most cases the founding and timing of these companies was the outcome of individually rational decisions made in a larger environment of irrational expectations surrounding the promise of a new technology. It was not the first time such a situation existed nor is it likely to be the last.

Moreover, a number of new and innovative companies such as eBay, Google, and Yahoo! emerged alongside the examples of sock-puppet mascots, and overly ambitious investments in infrastructure and web sites, that are the exclusive concentration of more reductionist analyses (Kaplan, 2002). While many companies lost significant amounts of capital (often spent on things that had little to do with improving the company's bottom line), others developed technologies and infrastructure which were later acquired by other companies allowing for a whole range of products and services that hitherto had not existed. Likewise, the experience gained by the entrepreneurs within these companies further built the skill sets within a region's economy to form companies in the future.

In short, while plenty of money was spent chasing bad business models, the impact of dot-com boom and bust has much more complex implications in the short and long terms. Even in the face of numerous bankruptcies, accounting scandals, and a weak economy, the dot-com era is not without its upside and mirrors the long history of dynamism and change within capitalist economies. Precisely because the public policy levers for supporting innovation-led economic growth are indirect (e.g., providing business environments that are conducive for experimentation and startups) rather than direct (picking winning technologies or companies), it is vital that policy-makers and citizens recognize that the dot-com era is not so much an anomaly but the most recent manifestation of Schumpeterian creative destruction.[1]

After the Downturn

While it is virtually certain that the 1990s will be used as an example of excess akin to the Dutch tulip craze of the 16th century, one needs to get past simplistic analyses of how much money was spent on tulip bulbs or dot-com stocks. Without doubt there was significant waste and many careless spending decisions were made in both cases, but one cannot halt the analysis at that point. After all, 350 years after the tulip craze the Netherlands remains a center for the tulip industry.[2] Dash (1999, pp. 215–16) explains that:

> The story of the tulip can be brought up to the present day in a very few words. The trade has continued to be dominated and driven forward by Dutch growers. Indeed, for much of the eighteenth century a single group of a dozen Haarlem florists effectively controlled the entire business. Even when their oligopoly was broken during the Napoleonic Wars, the reputation of Dutch farmers remained unparalleled, and as more and more people took up gardening as a hobby and worldwide demand for flowers of all sorts soared, the area around Haarlem given over to the cultivation of bulbs increased too.

The fortunes spent and lost on tulip bulbs are well documented but despite these individual losses, the regional economy around Haarlem and the Netherlands continued to be involved with tulip production for centuries. The resources were there, the knowledge was there, and the ability to market these flowers to an international market remained. In much the same way, regions central to the Internet industry (such as the San Francisco Bay area) have profited from the dot-com boom. This is by no means to belittle the loss suffered by individuals or the bankruptcies suffered by dot-com firms. Instead, the case study of the San Francisco Bay region places the rise of the dot-com firms in the context of the larger regional economy and highlights benefits that are external to firms and stock markets.

Stronger entrepreneurial climate

While profits, skills, and experience may escape the boundaries of firms, individuals retain them through personal human capital development, increased skills, or greater connectivity and social networking within the larger regional economy. Saxenian (1994) argues that the entrepreneurial vitality of the San Francisco Bay area is tied to the concentration of this connectivity and the consequent blurring of firm boundaries. Further, Saxenian (2000, p. 153) contends that:

> These networks promote new product development by encouraging specialization and allowing firms to spread the costs and risks associated with developing technology-intensive products. They spur the diffusion of new technologies by facilitating information exchange and joint problem solving between firms and even industries. Finally, the networks foster the application of new technologies because they encourage the entry of new firms and product experimentation.

Thus, while failed dot-com companies and their investors may no longer profit from these resources, they remain part of the region's resource base.

Expanding beyond inter-firm linkages, a number of actors active during the dot-com boom, e.g., specialty law firms (Suchman, 2000), venture capitalists (Kenney and Florida, 2000; Zook, 2001), industry trade groups (Saxenian, 1994), and ethnic or national affiliations (Saxenian, 1999), play key roles in supporting the entrepreneurial process. Regardless of the success or failure of the dot-com firms, the connections made by these actors during the boom can serve to support future companies and entrepreneurs.

Wasteful spending?

A common lament in the months following the slide of the public markets was the extent to which capital resources were "wasted" on dubious business modes and misplaced spending. When pressed, most of the founders of dot-com companies readily admit that money was not used well. Senior managers are happy to list the wastefulness of the spending patterns, although generally attribute most of it to other companies: "About 90 percent of the spending was wasteful. Some of the worst business decisions were made and remarkably little was invested in powerful technology or business models. Lots of it was spent in aimless tasks chasing bad business models." Others are even more critical, noting that:

> Absolutely money was wasted. We wasted money everywhere on just idiotic stuff. You give people a million bucks and they will spend a million bucks but if you tell them to do it for cheap they will do it. But few people had concerns about spending money then . . . I think that companies spent so much because of the market and the VCs who became so convinced that they couldn't fail that they just went, Go! Go! Go!

While not without a significant level of culpability, many dot-com executives were faced with public markets that were clearly eager to invest in their companies, providing them with access to relatively cheap money. Many founders recount that the relative ease with which money could be raised made it a relatively simple decision to pursue it. While the surplus of capital did lead to examples of wasteful spending, the emphasis during this time was on speed and growth. The founder of a failed Internet service firm argues:

> I don't think that too much capital is the reason we failed but I do think we had too much money. Once you get the money it has a momentum of its own. Things were moving so fast that the discipline on expenses was small because you needed to get stuff done without considering the bottom line. I'm not sure if you could have done anything about it. If you had not been getting big fast enough you probably would have been fired. If the market was willing to give you cash quickly, you take it. I don't want to be a revisionist and say how stupid we were because we were moving so fast. It was the way of the time. I don't think the lesson is to put less money in but to create an accountable framework for fast growth. After all resources do equate with success.

Other interviewees support this argument and point to the fact that the speed and the demands of the market made any other course difficult to secure. While certain actors were more closely tied to the raising of money, it is not entirely clear whether any one group could have had the power to slow this trend. Venture capitalists, investment banks, entrepreneurs, employees, and investors all had a common interest in maintaining the forward momentum. As a dot-com CEO notes:

> I don't know if you could have stopped it.. The institutional forces were all aligned around the outcome that we had. The only people who could have stopped it was an activist shareholder group but most shareholders were speculators and didn't give a rat's ass about the company as long as they could pass on their stock to the next person. The greater fool theory worked for longer than any one expected and as a result the hangover has been large.

The question of whether money was wasted turns on the process of technological innovation at the core of the Internet industry. The past century has seen numerous technological innovations – vacuum tubes, wireless, semiconductors, PCs, local and wide area networking, and fiberoptics – that were commercialized within the region. Many of the major companies that now populate the region (Intel, Apple, SGI, Genetech, Sun, etc.) were initially funded by venture capitalists willing to back companies on the basis of the promise of a new technology, a feasible business plan, and experienced entrepreneurs.

The emerging commercial Internet provided venture capitalists and other investors with an opportunity that looked as promising in terms of risk and reward as earlier rounds of technology had. Risk capital is so named precisely because of the high level of risk involved with backing young companies with new products that do not have a proven market. It is a game of numbers and venture capitalists were willing to risk the numbers precisely because for a time the promise of the Internet and dot-com companies was so compelling. A reflective venture capitalist contends:

> Money was not wasted because when the Internet broke there was enough promise there to second guess what we now know that we shouldn't have second guessed. Specifically the concept of this electronic global accessibility had implications that sounded pretty compelling. All of a sudden customers could access you at zero cost . . . unfortunately this proved to be wrong. We were willing to put capital at risk, to take the risk that these assumptions were correct. But this was Silicon Valley drinking its own bath water. We thought that the value

proposition we could create would be so compelling that people would change their behavior. With those beliefs you could then see creating these multibillion dollar companies and then the mindset was if we could do it, then anyone could do it, so we have to do it faster than anyone else. Then all this money starts showing up and saying here we'll pay you to get big fast and it fed on itself. That was the underlying logic that drove these pursuits.

Somewhere into 1999 it became clear that those assumptions one way or the other were fallacious. One of the reasons they were fallacious is that so many people started coming into the market that we started spending hundreds of millions of dollars in ads against each other. If there weren't all these entrants the fact of the matter is that more companies might have succeeded. It was kind of a lifeboat thing. Everyone jumped in so the whole lifeboat sank.

Looking forward from the standpoint of 1996, the potential of the Internet was compelling and already a number of companies, such as Netscape, Excite, Lycos, and Yahoo!, had shown that companies could emerge and go public based on this premise.

Nevertheless, it is troubling that no one, particularly venture capitalists, investment bankers, or policy-makers, had the backbone to stand up in response to the frenzy and suggest a more prudent course of action. However, it was in no one's interest or purview to do so. There were fortunes to be made, venture capitalists' venture funds were heavily oversubscribed, the Federal Reserve and Alan Greenspan were concerned with maintaining economic growth in the face of various international financial crises, and the public markets clearly were demanding dot-com stock almost regardless of the business plans put forth by companies. Many founders certainly had misgivings about the valuations of companies (although none were willing to say so on record) but many also argued that capitalism is about "Buy low, sell high" and there were plenty of willing buyers.

Survivors of the downturn

While many Internet companies have gone bankrupt, a number of dot-com firms have survived. Certainly not at the high valuations of the late 1990s but during the dot-com bust and the current recession, survival is a measure of success in and of itself. Many of the trappings of the dot-com era – Aeron chairs, razor scooters, postindustrial loft workspace, and a cool-sounding name – turned out to have much less traction than many expected. Nevertheless, as the CEO of a web portal company describes, it is possible to maintain Internet companies even in the current depressed and technology-cynical economy.

> The reason that we're still here is that we have a product strategy that creates value for our customers. It sounds simplistic but we're around because we deliver value to our customers which was rare in the dot-com world. It's not because we have the best funders, or the best resumés or necessarily the best technology but simply because we can deliver this value. All good businesses are eventually funded by their customers – not their investors – and it is just a matter how long it takes to get there.

The flush of money and the expectations of easy wealth downplayed this fact during the 1990s but it returned with a vengeance in the spring of 2000.

Money in and of itself, however, is not inherently a problem (although a generous supply of it certainly led to freer spending than would have otherwise occurred and even to misspending) and relatively cheap capital has proven instrumental in the survival of many dot-com companies. Paradoxically, the capital markets of 1999 and 2000 allowed some companies to amass a treasury that has seen them through the downturn and allowed them to experiment with new business models. A surviving dot-com firm's CEO explains how raising funds in 2000 has given them the flexibility to shift from an unviable business model to one that can work in the current market.

> I'm glad we raised money when we didn't need it and that we did cutting when we needed it. The markets were wide open at that time in early 2000 and the best time to raise money is when you don't need it . . . We would have been gone without the money [raised in the last round] by the first quarter 2001, but with it we were able to transform the company in 12 months and here we are.

This company's example is not unique as others have also been able to use the access to cheap capital during the boom as a point of leverage. This does not mean that a company with a bad business model

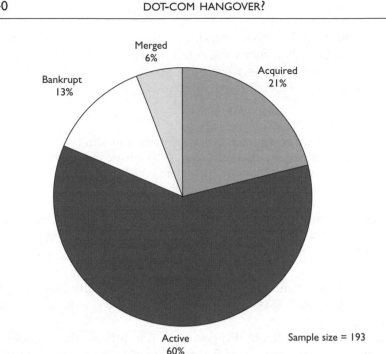

Figure 9.1 Status of public dot-com firms, San Francisco Bay region, June 2002.

which raised money during the time would necessarily continue to prosper. A number of companies who raised massive amounts of capital were unable to sustain themselves because of massive overheads (Webvan), bad business models (Kozmo), poor management decisions (Bigwords), or a combination of all three.

In addition, the extent to which companies have failed in the post-2000 world is generally overstated. While large bankruptcies are easy to identify, there are a number of companies that have been able to survive although not necessarily prosper during this time. Figures 9.1 and 9.2 show the fortunes of dot-com companies as of June 2002 that went public during the 1990s and 2000. While it is not a picture that investors were betting on during 1999, it shows that a relatively small number of outright bankruptcies took place. Moreover, while there are certainly companies that have not met the (now obviously unrealistic) expectations of investors, 60 percent of these companies are still active and independent, with a handful maintaining market capitalizations

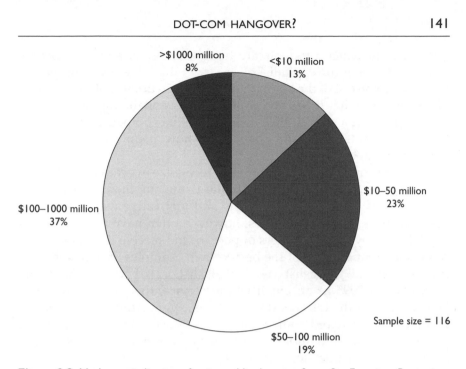

Figure 9.2 Market capitalization of active public dot-com firms, San Francisco Bay region, June 2002.

of greater than $1 billion and fully 64 percent with capitalizations greater than $50 million.[3]

There are also a handful that have emerged as clear winners within the economy, in terms of market capitalization or technological dominance. Yahoo!, eBay, and Google are all venture-backed companies that continue to operate and remain the dominant competitors within their respective market segments. By late 2004 eBay, Yahoo! and Google had market capitalizations of $73 billion, $52 billion and $50 billion respectively. Although these three companies only received a minuscule fraction of the approximately $80 billion invested in Internet companies located in the San Francisco Bay from 1995 to 2003, their market capitalizations alone can justify the amount of venture capital spent at the regional level.

This result is well in line with the structure of the venture capital investment model in which a very few companies pay for all other investments. And the big successes have introduced fundamentally new ways in which products are bought and sold and how people interact. As a venture capitalist observes, "I think there are companies that have come out of the boom that will remain dominant. A company like eBay and what it has accomplished by capitalizing on the Internet and making a shift in the economy. Media companies like Yahoo! have made quite dramatic changes in how people get news and entertainment."

One of the most important things to recognize in evaluating the dot-com period is that for every spectacular flame-out, there are examples of companies using the Web in new and innovative ways to expand their business. These companies, however, often have no formal risk capital, employ small numbers of people, and receive little in the way of media attention. One of the best-known examples of this is the listings site called Craigslist (www.craigslist.org). Founded by Craig Newmark in 1995 as an email listing service, this no-frills web site leverages what the Internet does best, i.e., aggregating relevant information from scattered sources in an easily accessible format. In the case of Craigslist, it catalogs things such as apartment listings, garage sales, and jobs listings for a local area.

While strongest in the San Francisco Bay area, Craigslist is beginning to be adopted elsewhere as a forum for sales including the notoriously byzantine rental market in New York City. As Newmark notes, the idea is simple.

> Craigslist is using the Internet to provide people with an opportunity to use it to do everyday stuff and make lives easier. Simple ordinary stuff like finding an apartment, getting a job or finding or selling an old sofa. I saw how the Well [an early Internet-based community] could provide a great virtual space to make real human connections and that's what we're trying to do with Craigslist. I'm not going to get rich but the development and deployment of the technology of the Internet during the 1990s provides the space for things like Craigslist which can ultimately be used to help solve human problems, big and small.

It is this simplicity and the low cost of use that makes Craigslist such a success and a marked contrast to the dot-com hoopla of the 1990s. It will not turn its employees into overnight millionaires but it does have the potential of growing steadily while providing sufficient revenues and profits to continue. In fact, because Craigslist did not go all out to capture a market, it has been able to expand in the post-2000 Internet.[4]

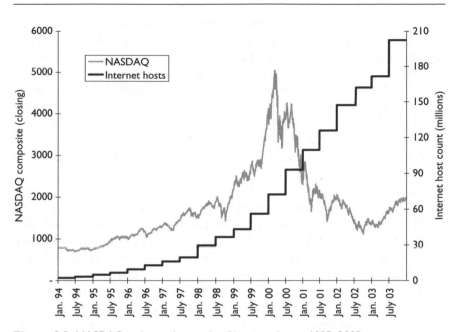

Figure 9.3 NASDAQ index and growth of Internet hosts, 1995–2003.
Source: NASDAQ and Internet Software Consortium/Network Wizards.

Global diffusion

In addition to the fortunes of individual companies, the dot-com boom had important implications for the adoption of the Internet by society. It is easy to forget that just 10 years ago Internet users numbered in the few millions while today the figure is over 600 million worldwide (NUA, 2002). Although the Internet would have attracted users with or without the dot-com boom, the publicity around the dot-coms, the plethora of free services they offered, and the allure of the cutting edge helped to create an increasingly large number of connections to the Internet (see figure 9.3).

Most importantly, despite the drop-off in stock indices like the NASDAQ composite, the number of Internet users continues to grow at a similar rate exhibited during the boom period.[5] Although dot-com founders might have overestimated the ability of their company to "change the world," the growth in Internet use is fundamentally changing the way in which people use and access data, communicate with one another, and view the world. For example, masked in the growth rate illustrated in figure 9.3 is the rapid growth of Internet use in China, providing considerable challenges to the current governmental system (Zittrain and Edelman, 2002a,b).

Although the dot-com boom is often equated with the tulip speculation in the Netherlands, an aspect shared by both booms is rarely considered, namely the ability of these booms to popularize the object upon which it had focused. As Dash (1999, p. 209) notes, "In an odd way the infamy that the mania had attracted helped too; the whole of Europe had heard of tulips now, and many people wanted to see for themselves the flower that had generated such passions." In a very similar fashion the dot-com mania of the 1990s cemented the idea of the Internet as part of everyday life in the minds of mainstream society.

Well before the start of the dot-com boom and bust, David (1990) showed how a pivotal technological innovation, the electrical motor first introduced in the 1880s, took four decades before altering the face of industry. It took that long for fundamental changes in the production process, e.g., transitioning from compact vertical factories to low-rise manufacturing able to take full advantage of this particular innovation, to be enacted. Likewise, firms and industries are at the very beginning of understanding how to use the Internet. From this knowledge, they gain an understanding of how this use will alter their geographies of production and distribution.

Prospects for Regional Economies

While the dot-com downturn has made finding work more difficult and slowed the migration patterns of the 1990s, it has also created some benefits. One of these is reducing the level of expectations cultivated during the boom phase. While the promise of a large payoff is an important part of the entrepreneurial process, the expectations had developed beyond a point of sustainability. As one company founder remembers:

> It got to the point that people had inflated expectations and felt that there was something wrong if they weren't a multi-millionaire after 15 months but that wasn't the norm here anyway. Now the culture inside companies is much more realistic. People came historically to Silicon Valley to change the technology world. I'm hoping we are getting that back.

In addition to a retrenchment of expectations within entrepreneurial companies, the bust period has served to discipline companies in a whole range of activities, from hiring decisions to marketing strategies. While entrepreneurs are by definition optimists, the influx of cash encouraged poorer financial decisions and shifted many companies' focus from controlling their burn rate to growing companies quickly at any cost. These attitudes have largely dropped away (or the companies where they continued to hold sway have gone bankrupt) and, as a venture capitalist observes, reset business culture attitudes.

> We've gone through a two year flushing period where a lot of the bad habits which were cultivated during the dot-com period have been flushed from the system. And I think the startups we are seeing now have the mindset in the business areas that are comparable if not better than they were in 1993.

As a result of more realistic expectations and better business (and investment) decisions, the dot-com companies remaining in the Bay area have a decided advantage vis-à-vis the late 1990s. Because so many companies were started, often in direct and fierce competition with one another, it was very difficult to attract employees, customers, and partners. The downturn has provided the surviving companies with easier access to employees and a chance to better develop their technology and business plans. A founder and CEO explains:

> I almost only see positive things going on now. Just the ability to gather a critical mass of people around a technology, a company. It wasn't possible in 2000. You couldn't rise above the noise. So that has been good for us. We're pleased about it. We could have been doing the same thing 24 months ago and how would anyone find out about us?

A final short-term benefit, although a truism, bears repeating, if for no reason than the overall focus on the short-term costs of the downturn. In short, economic downturns can offer a great boon to companies by reducing the cost of any number of factors of production, labor, land, and supplies. Those in a position to start companies are also able to benefit from this. As the founder of a retail Internet firm who is currently starting a specialized retailing company maintains, "I couldn't possibly start this store in a strong economy. We need a good location and the only way we could afford it was in this time. This is a great time to start a company because you can hire cheaply. You want to start a company in a weak economy." Of course, these lower costs often emerge as a result of someone else's loss, unemployed workers, real estate whose tenants are no longer in existence, but this can provide a spur in the short term for new entrepreneurial ventures.

Shifting the emphasis from financial to human capital

Although extremely difficult to measure at the regional level, a great deal of human capital and experience was gained during the boom period. Although not as easily quantifiable as academic degrees, people who worked at dot-com companies gained experience very quickly. While one could argue that the activities in which they were engaged were ill-thought-out and not particularly productive, the experience remains. This is particularly the case for the founders of many dot-com companies. As the cofounder of a surviving Internet firm reflects:

> On a personal basis I can't imagine an experience that would force me to scale the way that this has forced me to scale. I've done it now. I've managed people in a tough business situation with a business model that became tough to sell. I've had to lay off people and switch business models. I feel like I've had 10 years of business experience compressed into three.

Even in cases where companies have not survived, the experience of the founders can be long-lasting. It came at a high cost but provided individuals with a wide range of experience, such as described by the founder of a failed dot-com firm: "The good side is that I learned things that I never would have otherwise. I did things that I couldn't have dreamed of doing before it. That two years was worth 15–20 years of business experience." Labor, however, is mobile so there is no guarantee that the people who gained the experience will remain in the San Francisco Bay region. However, since a large part of the human capital is based in local and social networks within the region, it is possible that a number of these people will remain where their ability to maximize their experience is located.

Retooling venture-driven innovation

One of the ongoing questions concerning the Bay region's economy is the future of the venture capital-backed model of innovation. Developed over the course of decades, the venture investing system is an essential factor behind Silicon Valley's status as a premier technology region and has been emulated worldwide. The failure of so many high-flying venture-backed companies naturally leads to a questioning of whether the venture capital model itself is in crisis. While the late 1990s will no doubt be long remembered as a time when a great deal of risk capital was invested in less than stellar business ideas, it is not necessarily clear that the model will be fundamentally retooled.

Firstly, during the boom a number of individuals became active in venture capital investing without a great deal of experience and pursued companies with business models that have since proven very dubious. As these dot-com firms have failed these new venture capitalists will be unlikely to continue in this role. A former CEO argues:

> The venture capital model works when you have people who are extremely knowledgeable about an industry and are intimately involved in it. We got a lot of people working as venture capitalists who were not qualified but were doing it anyway. Fifty percent of these guys will not see all their committed capital committed. They'll do something else.

The late 1990s also saw a shift in the way established venture capitalists operated. Not only did venture capitalists provide less of the advice and connections as had historically been the case but early patterns of coinvestment broke down. A long tradition within venture capital is the syndication of investments by multiple venture firms as a way to lower risk and increase the support and social networks available to a new company (Timmons and Bygrave, 1986). However, as a Palo Alto-based venture capitalist contends, this strategy was often abandoned by investors during the dot-com boom period.

> What happened is that VCs had large funds and the path to liquidity was so quick and there was a tendency not to syndicate because it didn't look like it was needed. Because of that you didn't have the same partners around the table. Without syndication everyone backs their own company in the space and you have 50 companies in the market. Syndication gives more perspective and more bench strength, and more capital for play. The final thing in syndication is that there will be fewer "me too" investments. We could have used that in the 1990s. Building a good syndicate around one company means that it will be that much

tougher for the next company to build a similar syndicate and lead to
the kind of competition we saw between dot-coms.

While there is certainly room to question whether the online retailing of pet products, groceries, or beauty products would ever yield the
profits predicted by some, the fact that multiple venture capital firms
funded competing firms (often in direct response to another venture
capital firm) almost certainly guaranteed the level of cut-throat competition and multimillion dollar marketing campaigns that resulted
in dot-com firms spending tens of millions of dollars without any
noticeable results.

Altogether there is currently significantly less drive to invest in a
company at any cost and a greater willingness to consider syndication. This by no means suggests that venture capitalists are simply
happily cooperating with one another but reflects that the patterns of
the dot-com boom are more an aberration driven by the spectacular
financial returns than the typical way of conducting business.

Nevertheless, it is clear that venture capital investing has dropped
off sharply since the beginning of 2000 and has continued to steadily
decline quarter after quarter. However, this is a pattern within the
venture capital community that has been observed in the past. In particular, monies pledged to venture funds are not committed as limited
partners decide to cut back the share of their portfolio in risk capital.
It also reflects the higher standards and return to benchmarks instituted by venture capitalists after the heady days of the dot-com era.
A long-time venture capitalist observes:

> Some particular VC might be a little more gun-shy but in the spirit of
> Silicon Valley there is as much interest in taking risk. You could argue
> that there is even more than in 1995 in terms of the sheer number of
> people doing this stuff. One misinterpretation of things is that there are
> fewer startups getting their first round of financing. In comparison to
> the dot-com time people can say they must not be as risk tolerant as
> they were. But what we've done is gone back to the standards we had
> in 1993. A capital efficient approach to the business and an initiative that
> has a clear shot of being valuable to customers.
>
> We already have VCs dabbling in nanotechnology, which is an emerg-
> ing technology and where in some cases the market is five years out. I
> don't see it as an industry that is going wanting for risk capital. But that
> doesn't mean that there aren't thousands of nanotechnology want-to-be
> companies that are not getting backed because, guess what, there is not
> a business there. People are thinking that we're going to do what we

did in the dot-com period but we're not doing that anymore. We're doing what we've always done and evaluating the core business.

In the final analysis, steadily dropping venture capital investment is not so much a decline in the venture capital model as a return within the community to a focus on longer-term investments in new technology.

Living on the Edge of Novelty

The time has not yet arrived for a real summing-up . . . The gold rush is still on, and everything remains topsy-turvy. The analyst of California is like a navigator who is trying to chart a course in a storm: the instruments will not work; the landmarks are lost; and the maps make little sense. The last eight years have been, in fact, the most dynamic years in the history of this most dynamic state. No, the time has not come to strike a balance for the California enterprise. There is still too much commotion – too much noise and movement and turmoil. (McWilliams, 1949, p. 7)

This quotation in many ways captures the experience of the rise and decline of dot-com firms in the late 1990s, particularly in the San Francisco Bay region. McWilliams, however, was writing in 1949 in reference to the changes taking place during and immediately after the Second World War. These enduring themes of dynamism and change in California's economic landscape have been the few constants in a state that seems perpetually poised on what McWilliams terms the "edge of novelty." This historical parallel also provides an important perspective on the dot-com economy that unfolded in the San Francisco Bay region during the late 1990s. Although differing in the types of technology and business models deployed, this era mirrors earlier booms in the region and the state.

Risk, failure, and the regional economy

Although numerous dot-com companies have failed, this does not mean that the founders and managers of these companies need have a permanent black mark. It was well known (although for a period easily ignored) that startup companies are inherently risky ventures and those who get involved with them should be prepared for this risk. As one dot-com founder notes:

> The place [San Francisco Bay] appears to have a high tolerance for failure. It's almost worshiped as long as you don't screw your investors or employees or you don't do something immoral. If you have to fall, fall from a high place. You need a high level of risk acceptance and that is why I love this place and why it is the center of innovation.

The problem, of course, is that this fundamental fact was generally downplayed by all those concerned, founders, investors, and employees. This led to a higher degree of risk-taking by many individuals than they would normally consider, but the continual message was that "the greatest risk was not to be involved in the dot-com boom." This statement later proved hollow as it became increasingly clear that the greatest risk for people was the loss of money or livelihood through declines in stock price or bankruptcy of these companies.

Despite this risk and damage there remain a core of people who continue to engage in the risk and reward cycle of the Bay region. While certainly not enjoying the downturn and misfortune brought on by the bust, there remains an openness to risk-takers and new ideas within the region. Another long-time entrepreneur explains:

> There was no real fundamental change in what goes on here. Lots of people got caught up in the dot-com boom but had no idea of what they were into. It was all built on the boom and bust cycle but as an entrepreneur you should be aware of it. There is a mentality in California that makes it the most receptive place in the world for new ideas. I think there is a pioneer mentality. People have come from other parts of the world and the country. We still have the best access to capital, we have the smartest people in the world. Not just in Internet but biotech and everything else. It is simply more accepting of new ideas out here.

From this perspective the dot-com boom of the late 1990s is just another phase of the ongoing creative destruction of innovation in the Bay region. Firms are founded, grow, and disappear with great regularity in the region's economy. This process is aptly illustrated in figure 9.4, which is based on the *San Jose Mercury News* yearly reports on the largest 150 publicly traded firms in Silicon Valley (The Silicon Valley 150). This figure documents the number of new companies that have appeared on this ranking since 1995. This shows that while figures for 1999 and 2000 are peaks, they are not out of line with what regularly

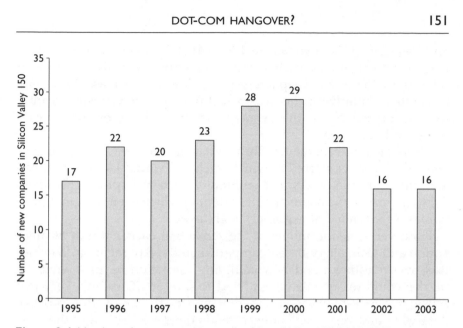

Figure 9.4 Number of new companies in the Silicon Valley 150, 1995–2003.
Source: Complied from the Silicon Valley 150 reports published annually by the *San Jose Mercury News*. For 2000, the reported number of newcomers was 34; however, in that year, new industries were included in the rankings for the first time. Excluding five companies that were included for the first time but would have ranked in the top 150 in previous years provides a more accurate representation of the number of new companies in the Silicon Valley 150 in 2000.

occurs within the rise and fall of companies within the region's economy. In short, while the dot-com boom contributed to an influx of large new companies during the 1990s and a subsequent exit of some of these companies post 2000, the dynamic this represents is a regular part of the San Francisco Bay's economy.

Excessive innovation?

Nevertheless, the extent of the excesses of the dot-com boom era and the impact of the subsequent downturn raises the question of whether it would have been possible to reduce the level of dot-com frenzy during the 1990s. In some ways, the dot-com era represents a unique time: the advent of discount brokerage services made it much easier for small individual investors to purchase dot-com companies' stocks and helped fuel the demand for IPOs. At the same time, large global macroeconomic issues, such as the Asian currency crisis, encouraged the Federal Reserve to keep interest rates low and increase the supply of capital. Underlying this was a technology that was significantly more approachable to the average person or investor than earlier waves of innovation.

These factors combined to form what in some ways may be considered a "perfect storm" in which each individual factor reinforced and amplified the others in a cumulative upward spiral. This likely led to the size of the boom and the extent to which even marginal companies were funded at extremely high levels.

However, it is not unique in the sense that earlier waves of innovation and technology-based companies in the Bay region also exhibited overinvestment and new company formation beyond what the market could reasonably be expected to support. Thus, in a very real sense, overinvestment is an inherent part of the innovation process. An earlier and very relevant example is the development of the hard disk drive industry during the early 1980s. From 1977 to 1984 over $400 million of venture capital was placed into 43 different manufacturers of Winchester disk drives and another $800 million raised in the public markets with results that convincingly parallel the fallout of the dot-com era 20 years later (Bygrave and Timmons, 1992).

Bygrave and Timmons (1992, p. 127) refer to this phenomenon as capital market myopia. They note that "Capital market myopia leads to the over funding of industries and unsustainable levels of valuation in the stock market. While we use the Winchester disk-drive industry to elucidate the phenomenon, capital market myopia has arisen in many other industries many times in the past. No doubt, it will occur in the future." Even more significantly, they observe that in addition to the downside of this myopia, consumers and the economy did achieve significant benefits. They note (1992, pp. 146–7), "Looking back on the industry, it is easy to overlook how consumers benefited from this competition. The cost of hard-disk drives today is a fraction of what it was even five or six years ago."

Moreover, it is impossible to know prior to an investment which companies and technologies will succeed. As Von Burg and Kenney (2000, p. 1152) argue:

> The difference between a radical innovation with massive capital gains and a mistake with no chance of success is not always easy to discern *a priori*. Many apparently sure things and great entrepreneurial visions

ultimately look foolish, because they find no customers, encounter problems that cannot be solved technically, or come to fruition only years or even decades after the first investments.

This point is particularly relevant for dot-com firms because of the great uncertainty in the 1990s surrounding the viability of business plans and how companies would turn a profit.

While recessions impact real people's lives in terms of unemployment, underemployment, reduced salaries, and general stress, they are part of a capitalist economy. The San Francisco Bay economy has gone through a number of downturns over time but often the most pressing issue at one moment is eclipsed by later developments. As one CEO remembers,

> I think it's important to keep things in perspective. For example, when I was in business school in 1985 all we talked about was Japan and how it was going to roll all over the world. They had this great system between the government and industry, they all worked incredibly hard. Now you look at the Japanese economy and wonder what we all were worried about. People just fixate on things and forget the size and complexity of the economy.

It is hopeful to reflect that the San Francisco Bay economy has proven remarkably resilient over time. While far from a guarantee for continuous growth, it is likely to provide a level of assurance that things may not be as dire as first conceived. When one lives on the "edge of novelty" it is impossible to play it safe and one never knows what indeed will be the new new thing.

Future implications

One of the greatest challenges facing regional innovative economies is recognizing the relatively small number of policy levers that can be successfully employed in shaping regional economies. While there are any number of necessary conditions for innovative growth, e.g., educated workforce, relatively easy technology transfer between university and commercial firms, service providers, and risk capital, these are not necessarily sufficient conditions. Moreover, it is not a straightforward process to create these conditions, as they emerge over time in the development process. As the dot-com experience shows, simply injecting money without the networks of know-who and know-how associated with it can lead to ill-advised investments and short-lived companies.

Although things are crystal clear with 20/20 hindsight, the real question is whether anyone at the time had the ability to put the brakes on the dot-com economy. A commonly repeated phrase throughout the interviews was "I drank the Kool-Aid," referring to the fact that despite plenty of indications of the flaws of the dot-com model, people were willing to suspend judgement because of the potential upside. The problem was that large swaths of the USA and the San Francisco Bay region were participating and "drinking the Kool-Aid." This is not to say that all participants benefited equally. There are numerous examples of founders and venture capital backers who were able to profit tremendously from the stratospheric rise in stock prices. More-over, there are a number of investment analysts, journalists, venture capitalists, and entrepreneurs who were touting dot-com stocks as the future of investments.

These activities certainly increased the demand for dot-com stocks, which ensured successful IPO after successful IPO and provided the demand signal that led to more venture capital investment. While these individuals are not guiltless, the real problem was that there was very little counter-narrative explaining why the emperor had no clothes. Although people would commonly joke about the high valuations of dot-com stocks, the expansion continued despite all expectations. One needs to remember that many people made a lot of money during the dot-com boom, providing them an interest in seeing it continue. A large number of people also lost a lot of money in the dot-com bomb, although those who made and lost are not necessarily the same people.

In short, while there were particular individuals who behaved unethically, it is too simple to lay the entirety of the blame for the boom and bust at their feet. As in the case of earlier waves of speculative frenzy such as the Dutch tulip market or Florida land speculation, the number of willing participants ranged much farther than a few individuals. Venture capitalists were funding dot-com firms but only continued to do so for as long as the public markets continued to demand dot-com stocks. The venture capital system operated in much the same way it had done before but was much more in the public eye than previously, leading to higher levels of participation. In short, a whole segment of the population was willing to accept the unrealistic expectations of the time. Clearly everyone needs to be more realistic but the nature of entrepreneurialism and innovation is based on defying realistic expectations.

Even if the lessons learned during the 1990s are adopted by this generation, there is no guarantee that a similar pattern will not occur in the future. Those who lived through the boom period will no doubt take the lessons to heart, but time has the ability to dull the sharpness of the lesson and ensure a continual supply of new entrants without the same level of personal experience with boom and bust. One need only look at history to see that speculative booms have been a steady refrain in economic development and are unlikely to disappear.

Entrepreneurs by definition are optimists about their ideas, their abilities, and their companies. Backers of these young companies need to spend significant amounts of time on due diligence but will continue to fund both wildly successful and wildly unsuccessful companies. In order to create new profitable companies and commercialize new technologies one needs also to support companies that will fail. It is the nature of capitalist development and an inherent part of the Californian economy since the gold rush of 1849.

In the final analysis there is no doubt that the birth of the Internet industry in the late 1990s will be remembered for its excesses, strange company names, and the tulip mania-like phenomenon that it represents. At the same time, however, it should be viewed as both the initial step that brought the wide commercialization and mass use of the Internet as well as yet another link in the ongoing history of technological development. The words of McWilliams (1949, p. 37) are once again relevant as well as prophetic: "In California you learn to wait for the next explosion and, when it comes, you run as far and as fast as you can and then dig in until the next explosion splits the air." Despite the fact that these words were written more than 50 years ago, they are relevant to the dot-com experience and describe much of what took place. The explosion of the commercial Internet led to a tremendous amount of activity, much of it wasteful, much of it in retrospect silly and ill-planned. The San Francisco Bay economy has now dug in, companies are getting by on as little spending as possible, and people are working less. Nevertheless, in research labs, startup facilities, and garages people are actively engaged in and working on creating the means for the next explosion that will split the air.

Appendix A:
Measuring the Internet Industry

Because of the Internet's decentralized infrastructure and organizational logic, locating either the consumption or the creation of its content is extremely difficult. Flows of data and communications can move around the globe with little regard for municipal, regional, or national boundaries. Despite this, information flows cannot exist without the people (living in physical space) who create, regulate, distribute, and consume the products and services generated by the Internet industry.

Researchers have relied on a number of different indicators to analyze the geography of the Internet, including Internet hosts (Hargittai, 1999; Jordan, 2001), bandwidth (Abramson, 2000; Townsend, 2001b; Malecki, 2002), IP addresses (Cheswick and Burch, 1998; Dodge and Shiode, 1998), links between web pages (Brunn and Dodge, 2001), domain names (Moss and Townsend, 1997; Kolko, 1999; Zook, 2000a,b, 2001), and lists of top web sites (Paltridge, 1997). While each provides insight on one aspect of the network, some are more appropriate for the analysis of the Internet industry. This appendix outlines data sources commonly used to map the Internet and details the methodology behind their generation.[1]

Hosts

A widely used indicator of growth and distribution are Internet hosts (Hargittai, 1999; United Nations Development Program, 1999; Internet Software Consortium, 2000). Although there is a great deal of variation between hosts, ranging from a single desktop computer to powerful servers acting as multiple "virtual" hosts, this measure gives a rough indicator of the minimum size of the Internet. While this provides a valuable metric of growth over time, it is not a straightforward process to assign these Internet hosts to geographic locations (OECD, 1998). This is further complicated by the fact that the majority of hosts are under gTLD domains, i.e. com, net, or org, and are not associated with any particular country.

A second and more important shortcoming to the use of hosts as an indicator for the geography of the Internet industry is the inclusiveness of the term. Because the definition of hosts encompasses both web servers offering web pages (supply) and PCs or ISP modems (demand), an important distinction between the production and use of Internet products and services is absent. Additionally, data on hosts are generally not available at any geographic level lower than the nation, which makes it difficult to identify domestic concentrations within the USA. Therefore, while host counts provide an indicator of the general size of the Internet, they are not a particularly useful indicator of commercial activity on the Internet.

Internet Infrastructure

Because the Internet is a network of networks using the TCP/IP protocol to exchange information, a number of researchers have mapped the geography of the Internet based on the structure of these networks (Cheswick and Burch, 1998). When this structure and the connections between Internet routers is analyzed, it is immediately possible to see the decentralized topography of the connections between Internet nodes.[2] This exercise, however, largely illuminates the technical topography of the Internet rather than the economic geography of the people and firms using it. The criterion for selecting the hosting of a web site, i.e., speed and reliability of connections, is very different from the factors that influence the location where a web site is created, i.e., access to skilled labor, capital, and amenities. It is entirely possible that a firm decides to host its content on a server farm located hundreds or thousands of miles from where it is designed and created.

A more geographically meaningful way of measuring the infrastructure of the Internet is the distribution of Internet bandwidth, i.e. telecommunications lines dedicated to Internet data packet traffic. As Abramson (2000) argues, in the absence of good measures of actual data traffic, bandwidth capacity between countries and cities provides a rough indicator of this traffic and the role of geography. For example, many Internet connections between computers in European or Asian countries are often first routed through the USA because of the greater bandwidth capabilities (TeleGeography, 2000). In his research on bandwidth between metropolitan areas, Townsend (2001b) shows that major cities play an important although not dominant role in the structuring of Internet bandwidth. While bandwidth is an important indicator of places capable of supporting an Internet industry, it does not guarantee that this will occur. Metropolitan areas with large bandwidth connections to other regions may act as digital transshipment points rather than suppliers of commercial Internet content.

Thus, while hosts and network infrastructure are useful in demonstrating the rapid expansion of the Internet, they are less successful when more specific topics about the supply and demand of web products or concentrations in subnational locations are addressed. Even the most relevant indicator, bandwidth capacity, cannot successfully distinguish between regions that are important dot-com industry locations and those that merely have the infrastructural potential to be such locations.

Domain Names

A better indicator for the location of the Internet industry is the registration addresses for domain names, such as nytimes.com or nokia.fi (Moss and Townsend, 1997).[3] Although registering a domain name has become relatively easy and inexpensive, it nevertheless represents a conscious decision to use the Internet in a more sophisticated manner. In many ways domain names are one of the most basic building blocks of the commercial Internet. Although actual data packets are routed by computers according to IP addresses, these numbers (e.g., 169.229.39.137) are hard for humans to remember. The domain name system was developed so that users could use an Internet address, for example www.zooknic.com, rather than its numeric equivalent.

Associated with each domain name is the unique contact information of the person or entity that registered it, which is available via an Internet utility known as Whois. As shown in figure A1 the data returned from a Whois query includes a billing address, contact names with phone numbers and emails, the date the domain name was registered, the last time it was updated, and the name servers respon-

Registrant:
Larry Page (GOOGLE-DOM)
  2400 E. Bayshore Parkway
  Mountain View, CA 94043
  US

Domain Name: GOOGLE.COM

Administrative Contact, Technical Contact, Billing Contact:
  DNS Admin (DA17675-OR) dns-admin@GOOGLE.COM
  Google, Inc.
  2400 E. Bayshore Pkwy
  Mountain View, CA 94043
  US
  650-318-0200
  Fax- 650-618-1499

Record last updated on 23-Aug-2000.
Record expires on 16-Sep-2009.
Record created on 15-Sep-1997.
Database last updated on 15-Mar-2001 17:12:14 EST.

Domain servers in listed order:

NS.GOOGLE.COM   209.185.108.134
NS2.GOOGLE.COM   209.185.108.135
HEDNS1.GOOGLE.COM   64.209.200.10

Figure A.1 Example of a Whois query.

sible for the domain. The dataset for gTLDs (i.e., com, net, org) used
in this book is based on automatic tabulations conducted every six
months from July 1998 to January 2004. Data for ccTLDs (e.g., .uk, .jp,
.fr) are based on statistics posted on each country code registry's home
page.

An analysis based on the CorpTech database shows a strong correlation between the location listed in Whois records and the actual
headquarters of a firm. For 84 percent of these firms within CorpTech,
the zip code obtained from its Whois registration matched the zip code
in the CorpTech database at the 3-digit level (roughly equivalent to a
geographic area the size of a small to mid-sized city) and 73 percent
of these firms match at the 5-digit zip code level (roughly equivalent
to a neighborhood within a city).

Although domain names are still a useful indicator of Internet activity, it must be acknowledged that the activity associated with par-

ticular domain names varies dramatically.[4] The domain name yahoo.com is certainly a much more important site for content on the web than zooknic.com. This weighting issue is resolved somewhat by the fact that major Internet content firms generally register multiple variations of their domain name both to protect their Internet brand and to allow differentiation between various products they offer. This gives additional weight to the most important Internet content firms and counterbalances the phenomenon of smaller and less-used domains. Nevertheless, the use of total number of domain names is a cruder indicator of the Internet industry than is desirable. In order to counteract this problem, other measures of the Internet industry based on specially developed datasets are also used.

Because there is no readily available source of historical data on the registration locations of Internet domain names, the creation date of domain names is used to determine how the location of domain name registrations has changed over time. It must be noted that there is a degree of fallacy in using the domain name data in this manner because the registration address obtained in July 1998 is not necessarily the same address at which the domain name was initially registered. However, given that firms value the maintenance of a consistent domain name identity, these data should provide a reasonable sense of how the location of Internet content production has shifted over time.

Top Web Sites

An alternative technique first discussed by Paltridge (1997) relies upon efforts on the Web to rank top web sites. This produces a weighted distribution of domain names that provides a better indication of the most important web sites. Although the exact methodology of these rankings systems are proprietary, they are generally based upon variables such as pageviews (number of times a site is accessed), unique visitors (counting individuals rather than hits), and other traffic measures. Three different sources of web-site rankings are outlined in table A1. While the results of any one ranking system should be interpreted with care, they do provide a sense of which are the most visited and used sites, and comparing the results of several "top site" rankings reduces the bias of any one methodology.

Of the three rankings shown in table A1, Alexa Research is particularly useful because it (1) provides the largest number of data

Table A1 Top web site rankings.

Name	Methodology
Alexa Research: top 1000 sites and pageviews	Number of pageviews from the aggregated traffic patterns of 500,000 web users worldwide
Media Metrix: top 500 sites	Unique visitors (multiple visits by the same person count only once) to web sites, ISPs, online services, e-commerce and other ad-supported sites
Go2Net: top 100 sites	Pageviews representing more than 100,000 surfers worldwide. Approximately 60% in North America and 40% elsewhere

Source: web pages of the ranking services.

points, (2) includes a weighting measure for individual web sites (i.e., number of pageviews), and (3) provides an ongoing time series. The domain names associated with these rankings can then be located geographically by using the registration information for the domain names that can be obtained from a Whois query.

Database of Internet Industry Firms

The final data source on the Internet industry is based on listings of Internet firms. Because governmental or other authoritative statistics on the Internet industry are generally not available at any geographic level smaller than the nation, regional data can only be created by aggregating firms available from various rankings and online databases. While these databases do not represent the entire population of Internet firms, they do provide hard numbers on employment, sales, and profits of the leading firms in the industry. This book uses three different sources of Internet industry firms that are outlined in table A2.

At the heart of the first database is Hoover's Online Business Network (http://www.hoovers.com/) that contains information on approximately 14,000 public and private firms worldwide. Firms were selected from this database if they were classified by Hoover's as belonging to the Internet sector or were otherwise identified by the author. While these 628 firms certainly do not include all companies

Table A2 Internet firm databases.

Name	Time series	Selection criteria
Hoover's Online: top 628 Internet firms	May 2000	Identified as Internet firms by Hoover's or selected by author
Interactive Week: top 500 companies in online revenue	November 2000	Ranked companies based on the amount of sales over the Internet, third quarter 1999 to second quarter 2000
National Retail Federation: top 100 e-tailers	1999	Top retail sellers from third quarter 1998 to second quarter 1999

Source: web pages of the ranking services.

in the Internet industry, they do represent a sample of the most important firms.

The Interactive Week and National Retail Federation (NRF) databases were constructed in a similar manner to the top web-site rankings. Each source ranks companies in terms of their amount of sales via the Internet and the geographic location for these firms is based upon the registration information for each web-site's domain name. Because each database was built using a different set of assumptions, e.g., NRF concentrates specifically on retail and Interactive Week looks at all sales (B2C, B2B) on the Internet, they are not directly comparable. Despite these differences their geographic distributions are remarkably similar.

Appendix B:
Interview Methodology and
Geographic Definitions

The use of interviews is essential in understanding the way the Internet industry has developed and the factors associated with its growth and decline. Using the ideas and techniques discussed by Hughes (1999), Clark (1998), Markusen (1994), and Schoenberger (1991), these interviews form a picture of the way this regional industry evolved.

Makeup of Interviewees

A total of 84 in-depth interviews were conducted with senior managers of dot-coms and early-stage venture capital investors in 1999 followed by an additional 53 interviews in 2001 and 2002. The majority of interviews were located in the San Francisco Bay region, with approximately 25 percent conducted in New York and 20 percent in Los Angeles. Eighteen of the people were interviewed in both periods. Interviews typically ranged between one and two hours and were all taped and transcribed.

Given the nature of some of the interviews, e.g., the reasons behind the failures of companies, and requirements placed on this research by the Human Subject Committee of the University of California, Berkeley, interview subjects were guaranteed anonymity. Therefore, statements are not directly attributed to any individual. In order to provide some background on the maker of a statement, brief identifications such as "founder of an e-commerce company in San Francisco" are used. A list of companies and venture capital firms interviewed (although not the individual interviewees) follows.

21VC Partners
24/7 Media
Absolute Reality
Again Technologies
AlleyCatNews
Angels Breakfast Club
Angels Forum
Artemis Ventures
AskJeeves.com
Autoweb.com
Babycenter.com
Bamboo.com
BayAngels
Bedrock Capital Partners
Beyond.com
Bidcom.com
Bigwords.com
Bikini.com
Bravata.com
Clarity Partners
Collabria.com
Colo.com
Craigslist.org
Cybersource.com
Dawntreader
Digital Chef
Doughnet
E-Greetings Network
Enterprise Partners
Epinions.com
ESS
E-Steel
Euclid Partners
Flatiron Partners
Frontier Ventures
Garage.com
Gay.com
GORP
Greenfield Technology Ventures
Guru.com
Hambrecht and Quist

Intelligenesis
Interloc/Alibrus
iPrint.com
Kline Hawkes & Co.
Listen.com
LivePerson
Looksmart.com
Mission Ventures
Moai Technologies
Mobius Venture Capital
Mpath
MyCFO.com
MyPoints.com
Netbuy
NetEarnings
Oasis Network Systems
Pacific Crest Securities
Palomar Ventures
Passlogix
Patricof Ventures
Paypal.com
Pets.com
Quokka Sports
Quova.com
Realnames.com
Redpoint Ventures
Reel.com
RocketVentures
Rolling Oaks Enterprises LLC
Small World Sports
Smart Technology Ventures
Snowball.com (IGN)
Softbank
Software Development Forum
Soliloquy
Sonnet Financial
SpringStreet
Storycatcher.com
Sutter Hill Venture
Tech Coast Angels
TheKnot

Hewson Group
Horizon Live
HungryMinds.com
iMotors.com
Information Technology
   Ventures
Instill.com
Institutional Venture Partners

Trident Ventures
Unicast
Virtual Vineyards
XIS (Novare)
Yaga (Xoom, NBCi)
Yahoo!

Over 70 percent of the senior managers of companies were also the founders of the companies and in all cases were in positions to reflect on the rise, fall, and/or success of their companies. Companies were selected from a range of sizes, products, visibility, business models, and status (operating, bankrupt, etc.) in order to span the gamut of experiences.

The interviews were unstructured but guided by the researcher and included questions on the following.

- Professional history of interview subject and his or her networks.
- Historical information on the company, its creation, and its fortunes.
- Overview of the company's business plan and strategy.
- Reflections on the dot-com boom and bust.
- Reason for shutdown or layoff (if applicable).
- Factors affecting the performance of the company.
- Employment status of the interview subject.
- Current and future career plans.

Analysis of the Interview Data

Each interview was taped and transcribed, often on the same day of the interview, but generally within three days, resulting in hundreds of transcribed pages. In every case, immediately after the interview, I identified and recorded three or four of the most important themes from the interview. The next step in the analysis of the data was analyzing the information provided by the subject in relation to both the issue that he or she was immediately addressing and the larger research questions of this research. Coding these paragraphs entailed the development of classification systems and making numerous judgments on borderline types and unclear responses. As a result the classifications were an evolving system throughout this analysis. Initial coding was followed by the process of looking for connections within the ideas and information presented by the interviewee but which had not necessarily been explicitly stated.

At the conclusion of the interview phase of the research, the interviews were analyzed as a group with the goal of identifying common themes from the coding done in the initial analysis. When a cross-cutting theme was identified, such as "We located here to be close to venture capital," the interviews containing this idea were compared with one another to identify linkages to other ideas that also might be held in common. This provides confirmation of data across interview subjects and a rich range of subtle differences, e.g., "We don't want to relocate to Colorado because we won't have the same connections to VC," "We can move faster because our VC is right here," or "I decided to move here from Toronto because there wasn't any venture capital there." At the same time, interviews containing a potential cross-cutting theme were examined for potential contradictory data such as "All of our customers are here." Cases containing contradictions were examined with greater scrutiny in the context of the entire interview to evaluate the weight given to individual statements, and again relied upon my informed judgment to assess the meaning.

Reliability and Quality of Data

The reliability of the data from interviews is of special concern and speaks directly to the confidence one can place in the arguments advanced by this book. For example, one must try to understand what a subject means by his or her sometimes contradictory responses, as well as the effort of people to present themselves in the best light. Because the companies and venture capitalists I interviewed relied heavily on public relations in their work, they wished that I received a positive impression of their company. The following quote from a San Francisco business-to-business company is typical of the way many of the Internet industry representatives characterized their companies.

> I'm passionate about what I do. This is about big business. This is about changing the world. It's exciting. We have clients that would amaze you if you saw what we were doing. I think it's one of the greatest opportunities in the world and that's not just a parent talking about his kid. I'm just talking pure market size and opportunity to get first movers advantage.

Nor were entrepreneurs shy about discussing the potential for the Internet and the activities surrounding the Internet industry. As the founder of an Internet company in New York remarks, "The mission of Internet companies is to fundamentally change the world." Another entrepreneur based in Redwood City, California, compares 1999 with the Italian Renaissance.

> The period of time in Florence known as the Renaissance was only 20 years and yet when you talk about the Renaissance period people think of that in terms of hundreds of years. I think we're in a renaissance period for Silicon Valley. It's not something that's happening across the nation, yet. Just like the Florence Renaissance period impacted all of Europe, so will Silicon Valley impact the rest of the world by what is happening here.

Venture capitalists were also self-promoting when discussing their own investments and investment strategies. Although at the time of many of the interviews firms were investing in companies with questionable business models, the interview subjects would invariably emphasize the quality of their own deal flow: "We're a value investor. We analyze each deal with its own metrics, sales, customers, these kind of long-term values. The current situation in venture capital, of investing in any idea is the opposite of what we do."

Initial interviews made it clear that asking directly about the issues that I was attempting to uncover would not provide me with much beyond the "public face" answers. In order to get at the more interesting but concealed knowledge, I developed a number of techniques. First, I emphasized my role as an academic researcher as opposed to a business press journalist and the fact that my research would not be published for several years. I also began each interview with general questions about the subject's career history that in addition to providing important data allowed me to establish a rapport with the subject before moving to potentially sensitive questions, such as "How long did you have to look for funding sources?" or "How many business plans did you send out before you got a meeting?"

Furthermore, I relied upon information gathered on the subject's career history to double-check the validity of data during the interview, e.g., "So you were still working at Cisco during the first three months of this company's existence?" Whenever possible, I asked specific questions based on information from secondary sources to explore topics that the subject might not bring up. For example, "How has your investment of $2 million in company X turned out?" While venture capitalists in particular were reluctant to talk about their bad investments, they were very willing to discuss the failings of other people's investments. This proved a good way of getting insight on how they go about sourcing and evaluating deals.

Although reliability of interview data remains an issue, it can also be a strength. For example, because I did not directly ask about the value of being near a venture capitalist (unless they had already mentioned it earlier) the meaningfulness of interview subjects mentioning this is greater than if they had simply checked a box in a survey. Additionally, the open interview approach allowed me to explore some issues in greater depth, e.g., the advantages and drawbacks of being located near to your venture capitalist.

Geographic Definition of the San Francisco Bay Region

The economic region of the San Francisco Bay used in this book is defined as the nine counties that comprise the San Francisco–Oakland–San Jose CMSA. These counties are Alameda, Contra Costa, Marin, Napa, San Francisco, San Mateo, Santa Clara, Solano, and Sonoma. The bulk of dot-com activity was concentrated in Santa Clara (Silicon Valley), San Mateo, San Francisco, and Alameda counties.

Notes

Chapter 1

1   A good example of this mindset is the contrast between eToys and Toys "R" Us in a 1999 *Business Week* article: "When old-economy companies have tried to beat their Net rivals at the new game, it has usually been the upstarts that prevailed . . . Perhaps nowhere will the contest between traditional and cyber-merchants be more intense than in toys . . . Toys 'R' Us Inc . . . is still struggling to get its cyberfooting . . . the entrenched E-tailer, eToys, will be hard to beat" (Zellner and Anderson, 1999, p. 31). In less than two years eToys was in bankruptcy while Toys "R" Us continued to expand its sales via the Internet.
2   Dot-com firms were nested in a larger technologic ecosystem of companies, e.g., Sun, Cisco, Qwest, Worldcom, Microsoft, providing much of the infrastructure and technology used. While encompassing some of the attributes of dot-com companies, e.g., Cisco's extensive online system of router sales, and clearly benefiting from the expansion of dot-coms, these companies did not depend on the Internet exclusively in their business models and in many cases were established well before 1994. For this reason they are considered as a separate but intricately connected set of companies distinct from dot-com firms.

Chapter 2

1   This history is intentionally brief. For a more thorough history of the development of ARPANET and the Internet, see Abbate (1999).
2   In addition, the acceptable use policies of the NSF were broadened. In June 1992, Representative Rick Boucher of Virginia introduced an amendment to the National Science Foundation Act of 1950 to allow for commercial uses of the Internet. Although a degree of commercial use of

NSFNET had previously existed, e.g., many of Cisco's router sales during the 1980s were transacted via email, this amendment was a crucial step in legalizing commercial activity on the Internet.

3   CERN was also connected quite early to the Internet although, as Abbate (1999, p. 94) notes, the original connection between ARPANET and CERN was through a somewhat illicit link via a computer at Cambridge University in England. This was due to the fact that the telecommunications company maintaining the link between CERN and Cambridge explicitly prohibited links between the ARPANET and CERN. Nevertheless, the development of the World Wide Web at CERN illustrates the long-reaching effects of this early, albeit underground, geography of the ARPANET.

4   The original program he wrote was called Enquire and allowed him to "store snippets of information, and to link related pieces together in any way. To find information, one progressed via the links from one sheet to another" (Berners-Lee, 1989).

5   As he notes in his proposal from March 1989, "A problem [at CERN], however, is the high turnover of people . . . information is constantly being lost . . . the technical details of past projects are sometimes lost forever, or only recovered after a detective investigation in an emergency. Often the information has been recorded, it just cannot be found" (Berners-Lee, 1989).

6   Constructing the browser and server, however, was only part of the solution to the larger issue taken on by Berners-Lee. Given the wide variety of machines in use at CERN and on the Internet, it was necessary to devise a method for ensuring that data was transmitted between servers and browsers in a standard manner. Just as Cerf and Kahn designed the TCP/IP protocol to negotiate data transfers between different types of networks, Berners-Lee designed a series of protocols for sending data via the World Wide Web. These include the Universal Resource Locator (URL) that provides the location of the data, the Hypertext Transport Protocol (HTTP) that defines how information is exchanged between computers, and the Hypertext Markup Language (HTML) that defines how data should be displayed on a computer (Naughton, 2000, p. 238).

7   Updates to these maps can be found at http://www.zooknic.com/users

8   Outside the USA, ccTLDs such as .uk and .de are often much more important than gTLDs such as .com, .net, .org. Therefore, when conducting international comparisons it is important to include both types to prevent the underemphasis of countries such as the UK and Germany and the inflation of Canada, where gTLD usage is higher than ccTLD usage.

Chapter 3

1   Because of the focus on the geography of the Internet industry, this analysis provides little on the geographies and spatialities of "virtual places."

See Dodge and Kitchin (2001a) for more information on this. Likewise it says little about the technical infrastructure of the Internet, a topic covered in Malecki (2002).

2 See Appendix A for details on the methods used to derive all these indicators.

3 The formula for the domain name specialization ratio is as follows:

$$\frac{\text{Number of .com domains in a region/number of firms in a region}}{\text{Number of .com domains in USA/number of firms in USA}}$$

4 Interestingly, such predictions were also made in the 19th century with the introduction of the telephone (Fischer, 1992).

Chapter 4

1 The author acknowledges that the scholars listed here represent a very broad range in the use of the word "institution."

2 The discounting of knowledge also reflects the difficulty in quantifying and modeling knowledge in the economy. As Romer (1994, p. 19) observes, "this kind of fact, like the fact about intra-industry trade or the fact that people make discoveries, does not come with an attached t-statistic."

3 Maskell (2001) points out that this is but one response that a firm may take and other options, such as out-sourcing, relocation, and automation, are also pursued. Scott (1996) also discusses the difference between "high road" and "low road" development strategies.

4 This overview is intentionally brief. See Foss and Knudsen (1996) for a more complete discussion of competence theory.

5 Von Hippel (1988) makes a similar but nonspatial argument around the interaction of producers and suppliers as a source of innovation and Porter, Enright, and Tenti (1990) highlight the importance of advanced users in spurring production.

6 Maskell uses this point to emphasize the problem of a purely transaction cost explanation, in the Williamson (1975) sense, for the clustering of firms. He argues that if firms clustered solely on the basis of transaction costs, then the most efficient structure would be a single firm in a region that eliminated all transaction costs. However, Maskell (2001, p. 8) asserts that such a structure would be inferior to a cluster of many firms due to "the specific forms of knowledge creation available to an individual firm when pursuing self-defined objectives, but not to a division of a larger entity where instructions are received and actions restrained by some procedure or limitation imposed from above."

7   This argument has come under challenge from researchers who argue that there is significant potential for codifying knowledge and that the magnitude of tacit knowledge has been exaggerated (Cowan, David, and Foray, 2000).

8   For an overview of the history and process of venture capital investing, see Bygrave and Timmons (1992).

9   This analysis is adjusted for inflation and is based on Venture Economics data, the only source that has time-series data that completely covers this period. However, the data cited here are limited to institutionalized sources of venture capital such as limited partnerships.

10 One of the most difficult tasks for any study of venture capital is determining an accurate count of the number of companies receiving venture investments, the sectoral definition of these companies, and the amount of these investments. Since venture capitalists are generally exempt from any governmental requirements on reporting their activities, it is necessary to rely on private sources, whose numbers can and do differ from one another and even change over time. The data used in this book come primarily from the PricewaterhouseCoopers/Venture Economics/ National Venture Capital Association (PwC/VE/NVCA) Moneytree survey circa 2004 (see figure 4.1).

11 Although during the Internet era these standards were not stringently enforced, Zider (1998) notes that historically firms needed "sales of about $15 million, assets of $10 million, and a reasonable profit history" to raise money in public markets.

12 While individual or angel investors are also important sources of risk capital (Gaston, 1989; Harrison and Mason, 1996; Helyar, 2000; van Osnabrugge and Robinson, 2000; Wetzel, 2003), lack of data on their investments means that they are notoriously hard to track. The line between angel and institutionalized investing is ill-defined and angels are increasing well organized and play much the same role in assisting companies that venture capitalists do. While differences do remain in terms of the amount of funding and abilities, in general this book includes angel investing when discussing the effects of venture capital investing.

Chapter 5

1   Saxenian's (1994, 1999) research has repeatedly documented this trend.

2   Ingham (1984) makes a similar observation on the difficulty that 19th-century British industrialists had in obtaining capital from London banks that were primarily focused on international trade.

3   Also problematic are the measures of the Internet industry used as the dependent variables in this model that are samples rather than complete populations of Internet firms.

4 The definition used is from Saxenian (1994), i.e., Standard Industrial Classification (SIC) codes: 357, computer and office equipment; 366, communications equipment; 367, electronic components and accessories; 376, guided missiles and space vehicles and parts; 38, instruments; 737, computer programming and data processing.

5 Information industries are defined by SIC codes as follows. Media and publication: 271, newspapers; 272, periodicals; 273, books; 483, radio and TV broadcast stations; 484, cable and other pay TV. Entertainment: 701, hotels; 781, motion picture production; 782, motion picture distribution; 783, motion picture theaters; 794, commercial sports; 799, miscellaneous amusement and recreational service. Advertising and public relations: 731, advertising; 874, management and public relations. Advanced users: 621, security brokers and dealers; 622, commodity contracts brokers; 623, security and commodity exchanges; 628, security and commodity services; 738, miscellaneous business services; 871, engineering and architectural services; 872, accounting, auditing, and bookkeeping; 873, research and testing services.

6 If these two regions were included in the analysis, it would make the findings even more supportive of the hypothesis of a causal role of venture capital investing and the location of the Internet industry.

7 While it is unusual to have completely orthogonal independent variables, one issue of concern for these regressions is multicollinearity among the independent variables. For example, the correlation between the log of venture capital investments and the log of employment is 0.47. While this correlation is high by some "rule of thumb" standards, this analysis includes the full range of variables in order to explore the full range of factors. Moreover, reduced models that dropped the employment and the location quotient for .com domains (the two variables most highly correlated with venture capital investment) remained predictive (r^2 0.53), with the venture capital variable significant at the 0.001 level.

8 An F-test on the full model (5) and the reduced model (4) is significant at the 99 percent level (df = 1, 92) and shows that one cannot reject that the variable of venture capital investing adds explanatory power to the model.

9 An F-test on the full model (5) and the reduced model (4) is significant at the 99 percent level (df = 1, 92) and shows that one cannot reject that the variable of venture capital investing adds explanatory power to the model.

Chapter 6

1 Florida and Kenney (1988b, p. 129) make a similar argument although without specific reference to managing knowledge. They argue that venture capitalists act as technology gatekeepers and represent a third avenue of innovation between Schumpeter's corporate and lone entrepreneur. They argue that venture capital

organizes the dynamic complementarities which exist among a variety of organizations, and as such represents a new, integrative model of innovation. In addition to this, venture capital-financed innovation plays an important technological gate keeping function – moving the U.S. across new technological frontiers and setting in motion the "gales of creative destruction" which establish the context for economic restructuring.

2 However, economic geographers have shown this connection. For example, Florida and Smith (1993, p. 448) state, "Geographic proximity functions to reduce uncertainty, compensate for ambiguous information, and minimize investment risk."

3 This screening method can be viewed as exclusionary, i.e., an old boys network, and some have accused venture capitalists of being overly clubby. This book, however, does not address this concern head-on since the focus here is on the mechanism used by venture capitalists rather than its ultimate effect on the equity of entrepreneurial opportunity. Nevertheless, this is an important topic for future research.

4 Coval and Moskowitz (2001, p. 811) document a related phenomenon in which they find that

> fund managers appear to earn substantial abnormal returns in their geographically proximate investments . . . Our findings suggest that fund managers are exploiting informational advantages in their selections of nearby stocks. Managers appear to earn abnormal returns in their local holdings as compensation for information they may acquire about local companies. This information may be the result of improved monitoring capabilities or access to private information of geographically proximate firms.

Chapter 7

1 For example, Cyrus McCormick received financing from the mayor of Chicago to market his reaper and Alexander Graham Bell's efforts to develop the telephone were financed by two wealthy Bostonians (Rosegrant and Lampe, 1992).

2 Walker (2001, p. 183) emphasizes the importance of the reinvestment of capital acquired through resource extraction.

> The third dimension of the spiraling circulation of capital in California was the rapid return of profits into new enterprise. This was developmental investment that went beyond resource grabs, rapid extraction, and self-aggrandizement. It marks a decisive moment of capitalism emergent and triumphant: using the wealth of nature as a lever to raise the level of productivity and widen the base of expansion.

Walker also emphasizes the distribution of nature's wealth to a relatively large petit bourgeois class in a process he characterizes as "prospector capitalism." Unfortunately, this book does not have the space to fully analyze the connections between this history of natural resource extraction and the origins of venture capital and will leave it for future research.

3 This includes the formation of one of the earliest wireless radio companies, Federal Telegraph. It was established in 1909 and backed by investments from the president and chair of the civil engineering department at Stanford, as well as a group of San Francisco investors (Sturgeon, 2000, pp. 19–20). A later example is the relocation of Philo Farnsworth, the inventor of electronic television, from Utah to San Francisco because of the willingness of his backer William Crocker, the son of a railroad magnate, to invest in his company (Sturgeon, 2000, pp. 34–5).

4 Although determining the first "true" venture capital investment in the San Francisco Bay turns on a number of definitional questions, Sturgeon (2000, p. 46) argues that Crocker deserves the honor. Another early example of an investment that is very similar to the current model of venture capital investing is the investment made by Henry McMicking in an audiotape company called Ampex. Introduced to the founder, Alexander Poniatoff, by Standfield Rayfield at Wells Fargo, McMicking invested $365,000 in the company for 50 percent ownership and assisted the company by bringing in a general manager and using his connections at the National Security Agency to lobby for contracts (Sturgeon, 2000, pp. 45–6). Kenney and Florida (2000, p. 105) argue that two venture investment companies, the Industrial Capital Corporation and the Pacific Coast Enterprises Corporation, both founded in 1946, "should be seen as West Coast precursors to venture capital."

5 It is likely that Kleiner's decision to seek investment for a new firm and Rock's decision to help was influenced by the successful IPOs of Varian Associates in 1956 and Hewlett Packard in 1957. These IPOs proved the market for small but fast-growing technology companies and also showed a pathway for companies besides acquisition by East Coast corporations (Kenney and Florida, 2000, p. 106). This idea of proving the viability of a sector's or region's companies reappears at the beginning of the commercial Internet era with Netscape's IPO in 1995 and in 1998 when DoubleClick's IPO "proved" the market potential for New York-based Internet firms.

6 Kenney and Florida (2000) argue that the Fairchild Semiconductor experience was also the impetus behind an innovation in contract and legal arrangements for venture financing. Law firms such as Wilson, Sonsini, Goodrich and Rosati pioneered contracts that provided key employees and founders with stock options that would provide incentives to stay with a company rather than the desire to leave that Fairchild Camera's purchase of Fairchild Semiconductors created.

7   Draper, Gaither & Anderson is generally credited with being the first California-based venture capital limited partnership (Wilson, 1985; Kenney and Florida, 2000).

8   The issue of compensation is probably one of the most important factors in the popularity of the limited partnership model. As the experience of SBIC shows, it is extremely difficult for other types of financial institutions to match the salaries that venture capitalists could make at a limited partnership. Kenney and Florida (2000, p. 113) report that, in particular, banks were unable to retain their managers once they had learned the ropes of venture investing. For example, during the 1970s the Bank of America lost seven people and Citicorp had lost 23 people to limited partnerships in Silicon Valley. A more recent trend has been the desire of accounting firms who both assist and audit startup companies to gain equity shares of companies, but because of regulatory and conflict of interest issues this is far from a straightforward process.

9   Some examples of this include Paul Wythes and William Draper III who founded Sutter Hill Ventures during the 1960s and became models of active venture investors with their focus on early-stage companies (Wilson, 1985). Other important firms included a partnership founded by Burton McMurty and Jack Melchor in 1969 called Palo Alto Investments, the Mayfield Fund formed as a partnership between Tommy Davis and Wally Davis in 1968, the Asset Management company founded by Franklin Johnson in 1965, and Capital Management Services (later renamed Sequoia Ventures), founded by Don Valentine in 1972.

10   For example, CNET relocated to San Francisco from New York in 1992 to be closer to this environment (Reid, 1997).

11   For example, in March 1993 only 0.1 percent of traffic on the NSFNET backbone was web based but in five months it had increased ten times and two years later it accounted for 24 percent of traffic (Naughton, 2000).

12   This prompted their Ph.D. advisor to suggest that they either shut down their site or move it off campus since the load was beginning to affect the university's computer network (Lardner, 1998). They temporarily moved the Yahoo! site from the trailer at Stanford to Netscape Communications. Although Netscape suggested that the two companies partner, Yang and Filo declined, although the relationship developed with Netscape during this time proved important to Yahoo! since Netscape was to later make yahoo.com the destination when a user selected the web directory via the Netscape browser (Reid, 1997).

13   By 1995, the Internet was becoming increasingly a topic of which the public was aware. Just a month before Netscape's IPO, *Time* magazine (June 3) ran a cover story on cyberporn based largely on a study from Carnegie-Mellon University that was of highly dubious quality. Nevertheless, the national spotlight offered by *Time* and the lurid nature of the topic brought a great deal of attention to the hitherto largely unknown Internet.

14 Yahoo! went public on April 12, 1996. The stock was priced at $13 a share but it opened at $24.50 and closed at $33.00 and was more successful than its competitor, Excite, whose own IPO had taken place a few weeks earlier.

15 This is using a restrictive definition of Internet companies that only includes content, business services, and e-commerce. A wider definition that would include ISPs, infrastructure, and software companies would place the amount at over $3 billion.

Chapter 8

1 One basic measure of the growth of these companies is number of .com domain names registered in the region. Although not a direct count of dot-com firms, it provides a useful proxy of the growth of activity focused on the commercial use of the Internet. From 2300 in January 1995, the number of .com domains registered in the San Francisco Bay area more than doubled every year for the rest of the century, with 73,000 in 1998 and 810,000 in January 2001.

2 Dot-com stocks were characterized by quick exchange, partly driven by day-traders and online brokers who had made it easy for large swaths of the public to buy and sell these stocks. For example, at the end of 1999 the average investor in DoubleClick held onto the stock for only five days (Byrne, 1999).

3 Although the amount of capital was large, the investing was concentrated in a relatively small number of firms. For example, during the 1998–2000 boom, $47.4 billon of venture capital was invested in 3754 rounds in 2924 companies. The number of venture-backed companies is remarkably small, on the order of 2–3 percent of all business within the region, especially given the level of media attention (according to US Census County Business Patterns, the San Francisco Bay region has approximately 100,000 business establishments).

4 At the start of 1998, Netscape began to show signs of strain from its competition with Microsoft in the browser market. Although Department of Justice antimonopoly proceedings had begun against Microsoft, the effects of this competition were taking its toll. In early 1998, Netscape reported an $88 million loss during the last quarter of 1997 and laid off over 400 workers (Hamm, 1998). Later in the year, the company announced its acquisition by America Online and in 1999, less than five years after it was founded, Netscape became a subsidiary of AOL.

5 Epinions later received a second round of $25 million in October 1999 and a third round of $12 million in February 2001.

6 These changes in focus of venture investing were reflected in the evolution of some entrepreneurs' business plans. As an Internet retailer based

in San Francisco notes, "I've seen some people follow the whole circuit. First they slapped on a .com to the end of their name and then they suddenly were selling things retail, then business to business, then with an auction. I don't know if they ever found backing but you could see them trying to join the trend of the month."

7   However, both companies would experience difficulties, e.g., PlanetRx went out of business in July 2000 and shares of Drugstore.com tumbled from highs in the mid-50s during September 1999 to around $1 a share in mid-2001, although it remains in business as of 2004. A similar situation existed with online retailing for pets, and three companies, Pets.com, Petopia, and Petstore.com, received more than $260 million in equity funding (by 2001 all were out of business).

8   It is interesting to note that the early history of the telephone in the USA saw a similar promotion of novel uses of a new technology. As Fischer (1992) notes:

> Telephone entrepreneurs in the early years broadcast news, concerts, church services, weather reports, and stores' sales announcements over their lines ... Telephone companies also offered sports results, train arrival times, wake-up calls, and night watchman call-ins. Industry journals publicized inventive uses of the telephone such as sales by telephone, get-out-the-vote campaigns, lullabies to put babies to sleep, and long-distance Christian Science healing.

9   In particular, John Doerr of Kleiner Perkins Caufield and Byers was a strong promoter of Internet companies and described the Internet as both "underhyped" and "the largest legal creation of wealth in the history of the planet" (Wylie, 1996). Doerr eventually apologized for this promotion and amended his statement to "the largest creation (and evaporation) of wealth in the history of the planet" (O'Brien, 2001).

10   Equally important to the process was the creation of demand for dot-com IPOs that provided a strong market signal for venture capitalists to invest in them. This demand was driven by positive analysis by many analysts, like Mary Meeker at Morgan Stanley and Henry Blodget at CIBC Oppenheimer and Merril Lynch. Equally important to this was the media, which was continuously reporting on these companies. As one senior venture capitalist notes:

> I don't think there are really any guilty parties but if I had an ax to grind I would pick on the media. The press portrayed that there was no risk. Nowhere was there any article saying, "You know what? A lot of people are losing their shirt." It got to the point where people were saying that the worst risk you can take is not to be involved. There was a gross impression

of tremendous rewards and no risk. But I think that the effect of the press was on the margin. At the end of the day this was a mass effort.

11 This first figure is based on data gathered from Hoover's Online Business Network (www.hoovers.com) in May 2000. The Hoover's database contained information on approximately 14,000 public and private firms worldwide. Firms were selected from this database if they were classified by Hoover's as belonging to the Internet sector or were identified by the author. From this list of 815 firms a subset of 628 firms were determined to have been founded explicitly to take advantage of the Internet. While these firms certainly do not include all companies in the Internet industry, they do represent a sample of the most important and leading firms in this industry. The Texas Internet Indicators study (June 2000) divides the Internet economy into four separate layers, ISP, applications, intermediaries, and Internet commerce. The final layer most closely corresponds to this book's definition of the Internet industry and is the source of the number cited above.

12 Disaggregated figures for the San Francisco Bay region are not available.

13 Although many founders of dot-com companies were able to create and, more importantly, maintain wealth, it is important to note that not all managers who participated in the gold rush have emerged with sizeable fortunes. Many individuals failed to achieve their fortune and actually found themselves at a point of net loss. As one dot-com CEO observes,

> Some people have had "less than zero" outcomes. There are a lot of executives who don't have a severance package but who took notes to buy company stock and now owe that to the company. Some people are suing for back wages and managers have to deal with that. Companies can have big funeral expenses.

14 One example of this is Toby Lenk, the founder of eToys, who was once worth close to $1 billion on paper but did not sell his holdings and as a result did not realize a financial gain (Sokolove, 2002).

Chapter 9

1 Economic history is replete with many instances where existing skills, firms, production processes, industries, and locations were confronted by a new innovation and in so doing offered the opportunity for new agglomerations and the reorganization of existing ones. As Walker (1995, p. 196) argues, "The amazing process of industrialization . . . has

repeatedly knocked the props out from under established social arrangements and posed new puzzles for humanity to solve. How this unwinds
is very much an open, experimental process, even though the contours
of the prevailing social relations channel the movements in certain ways."

2   Pavord (1999, p. 268) notes that "despite the way the tulip industry has
spread over the temperate areas of the world, the Dutch still remain
identified with the flower in a way that nobody else has ever quite
managed . . . The Netherlands exports at least two billion tulip bulbs a
year, two thirds of their total production . . . with an export trade worth
around £1330 million."

3   Additionally, there is evidence that the overall return from the venture
capital and public equity investments in dot-com companies had been
positive as late as the end of 2000. Using a nationwide database on publicly traded dot-com companies, Hendershott (2004) notes that at the end
of 2000 these companies were valued at more than the sum of all investments in them. Of course, valuations have dropped considerably in the
past four years but nevertheless this analysis suggests a stronger performance by dot-com companies than generally conceded. This, however, is
not a stronger performance by all dot-com companies but is due to the
spectacular performance of the most successful dot-coms, e.g., eBay or
Yahoo!, which started early on in the boom, 1995 or 1996, well before the
influx of capital.

4   Alexa Research reports that Craigslist was the 131st most visited web site
in the world as of September 2003. This compares with its ranking of 616
in December 1999, 409 in December 2000, and 256 in May 2002.

5   Other standard indicators of Internet usage such as e-commerce sales and
domain name counts have also exhibited continued growth.

Appendix A

1   In addition, Dodge and Kitchin (2001a,b) provide an excellent review of
the ways people have conceived of and mapped the Internet.

2   Another example is Malecki and Gorman's (2001) use of Internet backbone
data from Cooperative Association for Internet Data Analysis (CAIDA) to
construct a connectivity matrix for the USA and use trace routes to display
the connections between individual networks on the Internet.

3   Domain names are divided into two main types: gTLDs such as .com,
.net, and .org that can be registered by anyone; and ccTLDs which are
associated with particular countries, such as .uk for the United Kingdom
and .fi for Finland.

4   Another potential problem with domain names is the phenomenon of
domain name speculation, where a single individual registers hundreds

or even thousands of domain names in the hope of reselling them at higher prices. In order to counter this and produce geographically meaningful domain data, this database regularly identified and eliminated domain name holdings of 1000 or more.

References

Abbate, J. 1999. *Inventing the Internet*. Cambridge, MA: MIT Press.

Abramson, B.D. 2000. Internet globalization indicators. *Telecommunications Policy*, February, 69(6).

Acs, Z. and Audretsch, D. 1993. Innovation and technological change: the new learning. *Advances in the Study of Entrepreneurship, Innovation and Economic Growth*, 6: 109–42.

Alexa Research. 2000. *Internet Trends Report, 1999 Review* [Internet]. Alexa Research 2000 (cited 2000). Available from http://www.alexa.com/

Amin, A. 1999. An institutionalist perspective on regional economic development. *International Journal of Urban and Regional Research*, June, 365.

Amin, A. and Cohendet, P. 1999. Learning and adaptation in decentralised business networks. *Environment and Planning D: Society and Space*, 17(1): 87–104.

Amin, A. and Cohendet, P. 2000. Organisational learning and governance through embedded practices. *Journal of Management and Governance*, 4: 93–116.

Angel, D.P. 1989. The labor market for engineers in the US semiconductor industry. *Economic Geography*, April, 99(14).

Angel, D.P. and Engstrom, D. 1995. Manufacturing systems and technolgical change: the US personal computer industry. *Economic Geography*, January, 79–102.

Aoki, M. 2000a. Information and governance in the Silicon Valley model. In: X. Vives (ed.), *Corporate Governance: Theoretical and Empirical Perspectives*. Cambridge: Cambridge University Press, pp. 169–95.

Aoki, M. 2000b. *What are Institutions? How Should We Approach Them?* Working Paper 00015. Stanford, CA: Economics Department, Stanford University.

Aoyama, Y. 2001. The information society, Japanese style: corner stores as hubs for e-commerce access. In: T.R. Leinbach and S.D. Brunn (eds.), *Worlds of E-commerce: Economic, Geographical and Social Dimensions*. New York: John Wiley.

Arrow, K. 1962. The economic implications of learning by doing. *Review of Economic Studies*, 29(3): 155–73.

Bell, D. 1973. *The Coming of Post-industrial Society: A Venture in Social Forecasting*. New York: Basic Books.

Berners-Lee, T. 1989. *Information Management: A Proposal* [Internet]. CERN internal memo 1989 (cited 2001). Available from http://www.w3.org/History/1989/proposal.html/

Bronson, P. 1999. Instant company. *New York Times Magazine*, June 11, 44–7.

Brown, J.S. and Duguid, P. 2000. Mysteries of the region: knowledge dynamics in Silicon Valley. In: Chong-Moon Lee and W.F. Miller (eds.), *The Silicon Valley Edge*. Stanford, CA: Stanford University Press, pp. 16–39.

Brunn, S.D. and Dodge, M. 2001. Mapping the "worlds" of the World Wide Web. *American Behavioral Scientist*, June, 1717.

Bunnell, T.G. and Coe, N.M. 2001. Spaces and scales of innovation. *Progress in Human Geography*, 25(4): 569–89.

Bygrave, W.D. 1988. The structure of the investment networks of venture capital firms. *Journal of Business Venturing*, 3(2): 137–57.

Bygrave, W.D. and Timmons, J.A. 1992. *Venture Capital at the Crossroads*. Boston, MA: Harvard Business School Press.

Byrne, J. 1999. When capital gets antsy. *Business Week*, September 13, 72.

Cassidy, J. 2002. *Dot.con*. New York: Harper Collins.

Castells, M. 1996. *The Rise of the Network Society*. Cambridge, MA: Blackwell Publishers.

Castells, M. and Hall, P.G. 1994. *Technopoles of the World: The Making of Twenty-first-century Industrial Complexes*. London and New York: Routledge.

Cheswick, B. and Burch, H. 1998. *The Internet Mapping Project* [Internet] (cited 2001). Available from http://www.cs.bell-labs.com/who/ches/map/index.html

Clark, G. (1998) Stylized facts and close dialogue: methodology in economic geography. *Annals of the Association of American Geographers*, 88(1): 73–87.

Clark, G. and O'Conner, K. 1997. The informational content of financial products and the spatial structure of the global finance industry. In: K.R. Cox (ed.), *Spaces of Globalization: Reasserting the Power of the Local*. New York: Guilford Press, pp. 89–114.

Clark, J. and Edwards, O. 1999. *Netscape Time: The Making of the Billion-dollar Start-up That Took on Microsoft*. New York: St. Martin's Press.

Cohen, S.S. and Fields, G. 1999. Social capital and capital gains in Silicon Valley. *California Management Review*, Winter, 108(2).

Cortese, A. and Hof, R.D. 1995. Looking for the next Netscape. *Business Week*, (3447): 110–14.

Coval, J. and Moskowitz, T. 2001. The geography of investment: informed trading and asset prices. *Journal of Political Economy*, 109(4): 811.

Cowan, R., David, P., and Foray, D. 2000. The explicit economies of knowledge codification and tacitness. *Industrial and Corporate Change*, 9: 211–53.

Dash, M. 1999. *Tulipomania: The Story of the World's Most Coveted Flower and the Extraordinary Passions It Aroused*. New York: Crown Publishers.

David, P. 1990. The dynamo and the computer: an historical perspective on the modern productivity paradox. *American Economic Review*, 80(2): 355–61.

Davis, L. 1966. The capital markets and industrial concentration: the US and UK, a comparative study. *Economic History Review*, 19(2): 255–72.

Dodge, M. and Kitchin, R. 2001a. *Mapping Cyberspace*. London and New York: Routledge.

Dodge, M. and Kitchin, R. 2001b. *Atlas of Cyberspace*. London: Addison-Wesley.

Dodge, M. and Shiode, N. 1998. Where on earth is the Internet? An empirical investigation of the spatial patterns of Internet real-estate in relation to geospace in the United Kingdom. Paper read at Telecommunications and the City, March, Athens, GA.

Elango, B., Fried, V.H., Hisrich, R.D., and Polonchek, A. 1995. How venture capital firms differ. *Journal of Business Venturing* 10(2): 157–79.

Fenn, G., Liang, N., and Prowse, S. 1995. *The Economics of the Private Equity Market*. Washington, DC: Board of Governors of the Federal Reserve System.

Fischer, C.S. 1992. *America Calling: A Social History of the Telephone to 1940*. Berkeley, CA: University of California Press.

Florida, R. 1995. Toward the learning region. *Futures*, June, 527(10).

Florida, R. 2002. The economic geography of talent. *Annals of the Association of American Geographers*, 743–55.

Florida, R. and Kenney, M. 1988a. Venture capital high technology and regional development. *Regional Studies*, 22(1): 33–48.

Florida, R. and Kenney, M. 1988b. Venture capital-financed innovation and technological change in the USA. *Research Policy*, 17(3): 119–37.

Florida, R. and Kenney, M. 1988c. Venture capital and high technology entrepreneurship. *Journal of Business Venturing*, 3(4): 301–19.

Florida, R. and Smith, D.F. 1990. Venture capital, innovation and economic development. *Economic Development Quarterly*, 4(4): 345–60.

Florida, R. and Smith, D.F. 1993. Venture capital formation, investment, and regional industrialization. *Annals of the Association of American Geographers*, September, 434(18).

Foss, N.J. and Knudsen, C. 1996. *Towards a Competence Theory of the Firm*. London and New York: Routledge.

Freeman, J. 1999. Venture capital as an economy of time. In: R.T.A.J. Leenders and S.M. Gabbay (eds.), *Corporate Social Capital and Liability*. Boston, MA: Kluwer Academic, pp. 460–82.

Friedman, J.J. 1995. The effects of industrial structure and resources upon the distribution of fast-growing small firms among US urbanised areas. *Urban Studies*, June, 863(21).

Gaston, R.J. 1989. *Finding Private Venture Capital for Your Firm: A Complete Guide*. New York: John Wiley.

Gertler, M. 1984. Regional capital theory. *Progress in Human Geography*, XX: 51–81.

Gertler, M. 1995. "Being there": proximity, organization, and culture in the development and adoption of advanced manufacturing technologies. *Economic Geography*, January, 1(26).

Gertler, M. 2003. Tacit knowledge and the economic geography of context. *Journal of Economic Geography*, 3: 75–99.

Gilder, G. and Peters, T. 1995. City vs. country. *Forbes ASAP*, February 27, 56–61.

Gompers, P.A. and Lerner, J. 1999. *The Venture Capital Cycle*. Cambridge, MA: MIT Press.

Gorman, M. and Sahlman, W.A. 1989. What do venture capitalists do? *Journal of Business Venturing*, 4(4): 231–48.

Green, H., Himelstein, L., and Judge, P.C. 1998. Portal combat comes to the net. *Business Week*, (3567): 73–8.

Green, M.B. 1991. *Venture Capital: International Comparisons*. London: Routledge.

Green, M.B. and McNaughton, R.B. 1989. Interurban variation in venture capital investment characteristics. *Urban Studies*, April, 199(15).

Gupta, A.K. and Sapienza, H.J. 1992. Determinants of venture capital firms' preferences regarding the industry diversity and geographic scope of their investments. *Journal of Business Venturing*, 7(5): 347–62.

Hamm, S. 1998. The education of Marc Andreessen. *Business Week*, (3573): 84–92.

Hargittai, E. 1999. Weaving the western web: explaining differences in Internet connectivity among OECD countries. *Telecommunications Policy*, November–December, 701(1).

Harrison, B. 1992. Industrial districts: old wine in new bottles? *Regional Studies*, 26(5): 469–74.

Harrison, R.T. and Mason, C.M. 1996. Developing the informal venture capital market: a review of the Department of Trade and Industry's investment demonstration projects. *Regional Studies*, December, 765(7).

Hart, J.A., Reed, R.R., and Bar, F. 1992. The building of the Internet: implications for the future of broadband networks. *Telecommunications Policy*, November, 666(24).

Helyar, J. 2000. The venture capitalist next door. *Fortune*, 142(11): 292–312.

Hendershott, R.J. 2004. Net value: wealth creation (and destruction) during the Internet boom. *Journal of Corporate Finance*, 10(2): 281–99.

Herrigel, G. 1996. *Industrial Constructions: The Sources of German Industrial Power*. Cambridge: Cambridge University Press.

Himelstein, L. and Siklos, R. 1999. The rise and fall of an Internet star. *Business Week*, (3626): 88–94.

Howells, J. 2000. Knowledge, innovation and location. In: J.R. Bryson (ed.), *Knowledge, Space, Economy*. London and New York: Routledge, pp. 50–62.

Hughes, A. 1999. Constructing economic geographies from corporate interviews: insights from a cross-country comparison of retailer–supplier relationships. *Geoforum*, 30(4): 363–74.

Ingham, G. 1984. *Capitalism Divided? The City and Industry in British Social Development*. New York: Schocken Books.

Internet Software Consortium. 2000. *Internet Domain Survey FAQ* [Internet] (cited 2000). Available from http://www.isc.org/dsview.cgi?domainsurvey/faq.html

Jordan, T. 2001. Measuring the Internet: host counts versus business plans. *Information, Communication and Society*, 4(1): 34–53.

Kaplan, P.J. 2002. *F'd Companies: Spectacular Dot-com Flameouts*. New York: Simon and Schuster.

Kenney, M. and Florida, R. 2000. Venture capital in Silicon Valley: fueling new firm formation. In: M. Kenney (ed.), *Understanding Silicon Valley: The Anatomy of an Entrepreneurial Region*. Stanford, CA: Stanford University Press, pp. 98–123.

Kolko, J. 1999. The death of cities? The death of distance? Evidence from the geography of commercial Internet usage. Paper read at Cities in the Global Information Society: An International Perspective, November, Newcastle upon Tyne, UK.

Lardner, J. 1998. Search no further. *US News and World Report*, May 18, 49–53.

Leamer E.E. and Storper, M. 2001. The economic geography of the Internet age. *Journal of International Business Studies*, 32(4): 641–65.

Leinbach, T. and Amrhein, C. 1987. A geography of the venture capital industry in the US. *Professional Geographer*, 39(2): 146–58.

Lewis, M. 2000. *The New New Thing: A Silicon Valley Story*. New York: W.W. Norton.

Loasby, B. 1999. Industrial districts as knowledge communities. In: M. Bellet and C. L'Harmet (eds.), *Industry, Space, and Competition: The Contribution of Economists of the Past*. Cheltenham: Edward Elgar, pp. 70–85.

Locke, R. 1995. *Remaking the Italian Economy*. Ithaca, NY: Cornell University Press.

Lundvall, B. and Johnson, B. 1994. The learning economy. *Journal of Industry Studies*, 1(2): 23–42.

McWilliams, C. 1949. *California: The Great Exception*. Berkeley, CA: University of California Press.

Malecki, E. 1990. New firm formation in the USA: corporate structure, venture capital, and local environment. *Entrepreneurship and Regional Development*, 2: 247–65.

Malecki, E. 2000a. Knowledge and regional competitiveness. *Erdkunde*, 54(4): 334–51.

Malecki, E. 2000b. Creating and sustaining competitiveness: local knowledge and economic geography. In: J.R. Bryson (ed.), *Knowledge, Space, Economy*. London and New York: Routledge, pp. 103–19.

Malecki, E. 2002. The economic geography of the Internet's infrastructure. *Economic Geography*, 78(4): 399–424.

Malecki, E.J. and Gorman, S.P. 2001. Maybe the death of distance, but not the end of geography: the Internet as a network. In: T.R. Leinbach and S.D. Brunn (eds.), *Worlds of Electronic Commerce: Economic, Geographical and Social Dimensions*. New York: John Wiley, pp. 87–105.

Mandel, M.J. 2000. *The Coming Internet Depression: Why the High-tech Boom Will Go Bust, Why the Crash Will Be Worse Than You Think, and How to Prosper Afterwards*. New York: Basic Books.

Markusen, A. 1994. Studying regions by studying firms. *Professional Geographer*, 46(4): 477–92.

Markusen, A. 1999. Fuzzy concepts, scanty evidence, policy distance: the case for rigour and policy relevance in critical regional studies. *Regional Studies*, 33(9): 869–84.

Marshall, A. 1890. *Principles of Economics*. London and New York: Macmillan.

Martin, R. 1989. The growth and geographical anatomy of venture capitalism in the United Kingdom. *Regional Studies*, 23(5): 389–403.

Martin, R.L. 1999. The new economy geography of money. In: R.L. Martin (ed.), *Money and the Space Economy*. Chichester: John Wiley, pp. 3–28.

Maskell, P. 2001. Growth and territorial configuration of economic activity. Paper read at Nelson and Winter DRUID Conference, June 12–15, Aalborg, Denmark. Available from http://www.druid.dk/conferences/nw/paper1/maskell.pdf

Maskell, P. and Malmberg, A. 1999. Localised learning and industrial competitiveness. *Cambridge Journal of Economics*, March, 167(1).

Mason, C. and Harrison, R. 1994. The informal venture capital market in the UK. In: A. Hughes and D.J. Storey (eds.), *Finance and the Small Firm*. London and New York: Routledge, pp. 64–111.

Mason, C. and Harrison, R. 1999. Financing entrepreneurship: venture capital and regional development. In: R.L. Martin (ed.), *Money and the Space Economy*. Chichester: John Wiley, pp. 157–83.

Moschovitis, C.J.P. 1999. *History of the Internet: A Chronology, 1843 to the Present*. Santa Barbara, CA: Abc-Clio.

Moss, M.L. and Townsend, A. 1997. Tracking the net: using domain names to measure the growth of the Internet in US cities. *Journal of Urban Technology*, 4(3): 47–60.

Naughton, J. 2000. *A Brief History of the Future: From Radio Days to Internet Years in a Lifetime*. Woodstock, NY: Overlook Press.

Negroponte, N. 1995. *Being Digital*. New York: Alfred A. Knopf.

Negroponte, N. 1999. Being rural. *Wired Magazine*, issue 7.06, June, 94.

Nocera, J. 1999. Do you believe? How Yahoo! became a blue chip. *Fortune*, 139(11): 76–92.

Nonaka, I. and Takeuchi, H. 1995. *The Knowledge-creating Company: How Japanese Companies Create the Dynamics of Innovation*. New York: Oxford University Press.

Norberg, A.L., O'Neill, J.E., and Freedman, K.J. 1996. *Transforming Computer Technology: Information Processing for the Pentagon, 1962–1986*. Baltimore, MD: Johns Hopkins University Press.

NUA. 2002. *How Many Online*. NUA Internet 2002 (cited 2002). Available from http://www.nua.ie/surveys/how_many_online/index.html

O'Brien, T. 2001. Top VC Doerr apologizes for helping fuel dot-com frenzy. *San Jose Mercury News*, July 16.

OECD. 1997. *Webcasting and Convergence: Policy Implications*. Paris: OECD Publications.

OECD. 1998. *Internet Traffic Exchange: Developments and Policy*. Paris: OECD Publications.

Paltridge, S. 1997. *Internet Domain Names: Allocation Policies*. Paris: OECD Publications.

Pavord, A. 1999. *The Tulip*. New York and London: Bloomsbury.

Piore, M.J. and Sabel, C.F. 1984. *The Second Industrial Divide: Possibilities for Prosperity*. New York: Basic Books.

Polanyi, M. 1958. *Personal Knowledge: Towards a Post-critical Philosophy*. Chicago: University of Chicago Press.

Polanyi, M. 1967. *The Tacit Dimension*. London: Routledge and Kegan Paul.

Porter, M.E., Enright, M.J., and Tenti, P. 1990. The competitive advantage of nations. *Harvard Business Review*, March–April, 73(21).

Porter, M.F. 1998. Clusters and the new economics of competition. *Harvard Business Review*, November–December, 77(1).

Pred, A.R. 1977. *City Systems in Advanced Economics: Past Growth, Present Processes, and Future Development Options*. New York: John Wiley.

Putnam, R. 1994. *Making Democracy Work: Civic Traditions in Modern Italy*. Princeton, NJ: Princeton University Press.

Quarterman, J. 1997. Is ".com" primarily US or international? *Matrix News*, 7(January): 8–10.

Reid, R. 1997. *Architects of the Web: 1,000 Days That Built the Future of Business*. New York: John Wiley.

Rimmer, P.J. and Morris-Suzuki, T. 1999. The Japanese Internet: visionaries and virtual democracy. *Environment and Planning A*, 31(7): 1189–206.

Romer, P. 1994. The origins of endogenous growth. *Journal of Economic Perspectives*, 8(1): 3–22.

Rosegrant, S. and Lampe, D. 1992. *Route 128: Lessons from Boston's High-tech Community*. New York: Basic Books.

Sapienza, H.J. 1992. When do venture capitalists add value? *Journal of Business Venturing*, 7(1): 9–27.

Saxenian, A. 1994. *Regional Advantage: Culture and Competition in Silicon Valley and Route 128*. Cambridge, MA: Harvard University Press.

Saxenian, A. 1999. *Silicon Valley's New Immigrant Entrepreneurs*. San Francisco, CA: Public Policy Institute of California.

Saxenian, A. 2000. The origins and dynamics of production networks in Silicon Valley. In: M. Kenney (ed.), *Understanding Silicon Valley: The Anatomy of an Entrepreneurial Region*. Stanford, CA: Stanford University Press, pp. 141–64.

Schoenberger, E. 1988. Multinational corporations and the new international division of labor: a critical appraisal. *International Regional Science Review*, 11: 105–20.

Schoenberger, E. 1991. The corporate interview as a research method in economic geography. *Professional Geographer*, 43(2): 180–9.

Schumpeter, J.A. 1939. *Business Cycles: A Theoretical, Historical, and Statistical Analysis of the Capitalist Process*. New York and London: McGraw-Hill.

Schumpeter, J.A. 1942. *Capitalism, Socialism, and Democracy*. New York: Harper.

Scott, A.J. 1988a. *Metropolis: From the Division of Labor to Urban Form*. Berkeley, CA: University of California Press.

Scott, A.J. 1988b. *New Industrial Spaces: Flexible Production Organization and Regional Development in North America and Western Europe*. London: Pion.

Scott, A.J. 1996. Economic decline and regeneration in a regional manufacturing complex: southern California's household furniture industry. *Entrepreneurship and Regional Development*: 75–98.

Segaller, S. 1998. *Nerds 2.0.1: A Brief History of the Internet*. New York: TV Books.

Sokolove, M. 2002. How to lose $850 million – and not really care. *The New York Times Magazine*, 9 June: 64–68.

Sparks, D. and Laderman, J.M. 1999. The great Net stock sell-off. *Business Week*, (3642): 32–3.

Storper, M. 1997. *The Regional World: Territorial Development in a Global Economy*. New York: Guilford Press.

Storper, M. and Harrison, B. 1991. Flexibility, hierarchy and regional development: the changing structure of industrial production systems and their forms of governance in the 1990s. *Research Policy*, 20(5): 407–22.

Storper, M. and Scott, A. 1993. *The Wealth of Regions: Market Forces and Policy Implications in Local and Global Context*. Lewis Center for Regional Policy Studies, University of California, Los Angeles, Working Paper No. 7.

Storper, M. and Walker, R. 1989. *The Capitalist Imperative: Territory, Technology, and Industrial Growth*. Oxford: Basil Blackwell.

Stross, R.E. 2000. *eBoys: The First Inside Account of Venture Capitalists at Work*. New York: Crown Business.

Sturgeon, T. 2000. How Silicon Valley came to be. In: M. Kenney (ed.), *Understanding Silicon Valley: The Anatomy of an Entrepreneurial Region*. Stanford, CA: Stanford University Press, pp. 15–47.

Suchman, M. 2000. Dealmakers and counselors: law firms as intermediaries in the development of Silcion Valley. In: M. Kenney (ed.), *Understanding Silicon Valley: The Anatomy of an Entrepreneurial Region*. Stanford, CA: Stanford University Press, pp. 71–97.

Teece, D.J. 1998. Capturing value from knowledge assets: the new economy, markets for know-how, and intangible assets. *California Management Review*, Spring, 55(25).

Telegeography. 2000. *Hubs and Spokes: A Telegeography Internet Reader*. Washington, DC: Telegeography Inc.

Thompson, C. 1989. The geography of venture capital. *Progress in Human Geography*, 13(1): 63–97.

Timmons, J.A. and Bygrave, W.D. 1986. Venture capital's role in financing innovation for economic growth. *Journal of Business Venturing*, 1(2): 161–76.

Touraine, A. 1971. *The Post-industrial Society. Tomorrow's Social History: Classes, Conflicts and Culture in the Programmed Society*. New York: Random House.

Townsend, A.M. 2001a. Network cities and the global structure of the Internet. *American Behavioral Scientist*, June, 1697.

Townsend, A.M. 2001b. The Internet and the rise of the new network cities, 1969–1999. *Environment and Planning B: Planning and Design*, 28(1): 39–58.

United Nations Development Program. 1999. *Human Development Report 1999*. New York: United Nations.

Van Osnabrugge, M. and Robinson, R.J. 2000. *Angel Investing. Matching Startup Funds with Startup Companies: The Guide for Entrepreneurs, Individual Investors, and Venture Capitalists*. San Francisco, CA: Jossey-Bass.

von Burg, U. and Kenney, M. 2000. Venture capital and the birth of the local area networking industry. *Research Policy*, 29(9): 1135–55.

von Hippel, E. 1988. Trading trade secrets. *Technology Review*, March, 58–64.

von Hippel, E. 1994. "Sticky information" and the locus of problem solving: implications for innovation. *Management Science*, 40(4): 429–39.

Walker, R.A. 1995. Regulation and flexible specialization as theories of capitalist development. In: H. Ligget and D. Perry (eds.), *Spatial Practices: Critical Explorations in Social/Spatial Theory*. Thousand Oaks, CA: Sage, pp. 167–208.

Walker, R.A. 2001. California's golden road to riches: natural resources and regional capitalism, 1848–1940. *Annals of the Association of American Geographers*, March, 167.

Wetzel, W. 2003. Venture capital. In: W.D. Bygrave (ed.), *The Portable MBA in Entrepreneurship*, 3rd edn. New York: John Wiley, pp. 167–96.

Williamson, O. 1975. *Markets and Hierarchies, Analysis and Antitrust Implications*. New York: Free Press.

Wilson, J.W. 1985. *The New Venturers: Inside the High-stakes World of Venture Capital*. Reading, MA: Addison-Wesley.

Wylie, M. 1996. *Market Mover*. CNET NEWS.COM, September 10. Available from http://news.com.com/2009-1082-233580.html

Zakon, R. 1999. *Hobbes' Internet Timeline v4.0* (cited 1999). Available from http://info.isoc.org/guest/zakon/Internet/History/HIT.html

Zellner, W. and Anderson, S. 1999. The big guys go online. *Business Week*, September 6: 30–2.

Zider, B.O.B. 1998. How venture capital works. *Harvard Business Review*, November–December, 131(1).

Zittrain, J. and Edelman, B. 2002a. *Documentation of Internet Filtering in Saudi Arabia*. Working Paper, Berkman Center for Internet and Society, Harvard Law School. Available from http://cyber.law.harvard.edu/filtering/saudi-arabia/

Zittrain, J. and Edelman, B. 2002b. *Documentation of Internet Filtering in China*. Working Paper, Berkman Center for Internet and Society, Harvard Law School. Available from http://cyber.law.harvard.edu/filtering/china/

Zook, M. 2000a. The web of production: the economic geography of commercial Internet content production in the United States. *Environment and Planning A*, 32:411–26.

Zook, M. 2000b. Internet metrics: using host and domain counts to map the Internet. *Telecommunications Policy*, July–August, 613(8).

Zook, M. 2001. Old hierarchies or new networks of centrality? The global geography of the Internet content market. *American Behavioral Scientist*, June, 1679.

Index

Page references in *italics* denote tables, maps, or figures.

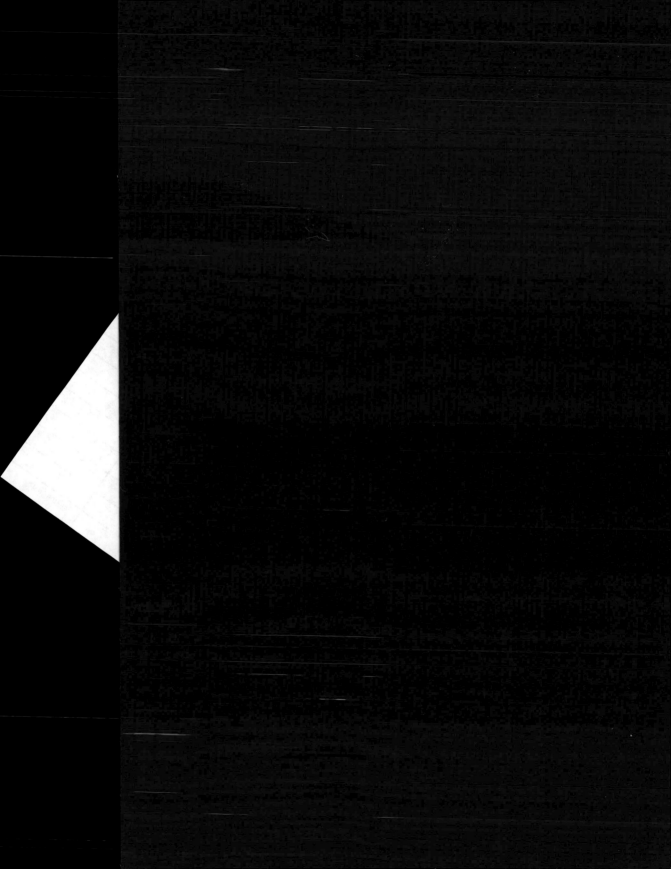

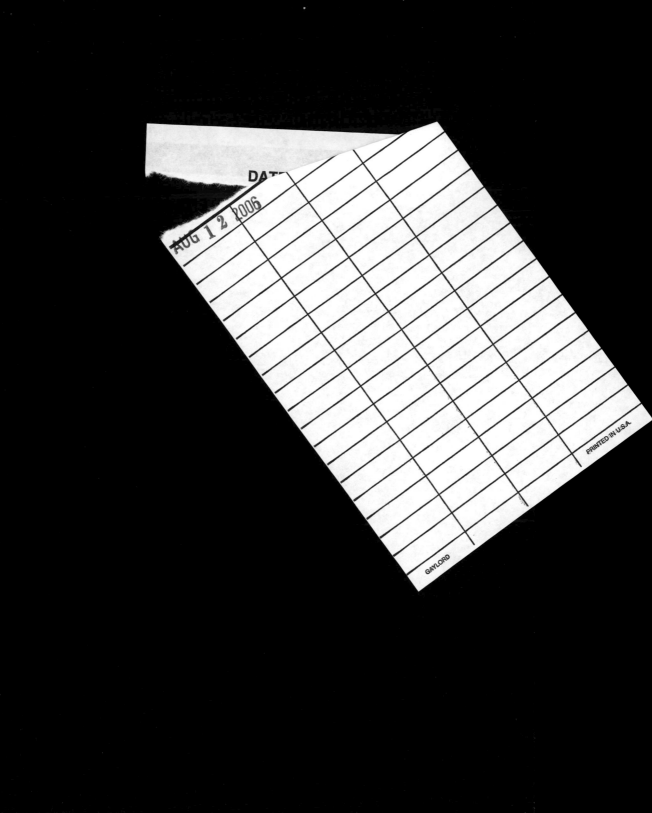